Desert
Plants
and
People

Desert Plants and People

by
SAM HICKS

The Naylor Company
Book Publishers of the Southwest
San Antonio, Texas

1st Printing 1966
2nd Printing 1971

Copyright ©, 1966 by THE NAYLOR COMPANY

This book or parts thereof may not be reproduced without written permission of the publisher, except for customary privileges extended to the press and other reviewing agencies.

Library of Congress Catalog Card No. 66-20096

ALL RIGHTS RESERVED

Printed in the United States of America

ISBN 0-8111-0041-3

To my wife, Ruby, who approved wholeheartedly of all these doings, and whose kitchen still harbors the faint but pleasant aromas left from the brewing of herb teas, I lovingly dedicate this book.

S.H.

Foreword

Sam Hicks, the author of this book, has been intimately associated with me for nearly twenty years. He is about the best all-around outdoorsman that I have ever met.

I first met Sam in Wyoming when he was working a ranch of his own, and working during the hunting season as a guide and outfitter.

Even Sam's competitors admitted that Sam was just about the best hunter in the state. Sam can look at an elk's track, follow it for a hundred yards, and then quit following the track and take you to the elk.

This sounds incredible, but I have had him do it with me more than once.

Like all tricks, there is an explanation. After Sam has followed the track for a hundred yards or so, he knows whether the elk is feeding, going to water, or going in to bed down. And Sam knows the country in which he hunts so intimately that he knows just about where the elk will go to feed, where it will go to water, and where it will go to bed down.

After I had been on two or three hunting trips with Sam, I knew that I needed him in my business; so I made it worth his while to give up his ranch in Wyoming, quit his outfitting and come down to southern California to live and adventure with me.

Together, we have been on many adventurous trips — down in Baja California, out in the desert searching for lost mines, using helicopters to explore terrain that has never known a footprint in modern times, and exploring the Delta country of the Sacramento River by houseboat.

Even before Sam knew me, he was doing writing on his own. And, from time to time, when Sam becomes interested in something, he writes about it, and writes about it entertainingly.

Sam is a gregarious soul who loves people, and when we went down into Baja California and Sam found there was a language barrier which kept him from understanding the people with whom we were camping, he started doing something about it — not by conventional study but by simply talking and listening, the way a child learns language, until he became quite fluent in Spanish.

And just as Sam likes people, so people like Sam. It wasn't long until the Mexicans and the Indians were telling Sam their innermost secrets.

And since many of those secrets had to do with herb remedies, Sam was duly impressed.

My ranch at Temecula is bounded on two sides by an Indian reservation, and Sam became acquainted with the Indians and their medicine and saw some very remarkable cures made by the use of herbs. So it was only natural that Sam started studying the herbs and then writing about them.

Much of this knowledge is rapidly disappearing from human ken — which is a shame, because the people who have this knowledge are the older people who can't be with us forever.

That is why I think this book in which Sam has collected so much of the information he has acquired is a valuable contribution to our knowledge as well as a valuable contribution to health.

I can't guarantee that the herbs have the medicinal

value which is claimed for them, but I can state that on my ranch I have a Mexican who believes in herb remedies. He has been with us for several years and during several of the so-called periods of "flu" epidemics.

When I start getting the flu, I don't monkey with it. I go to bed. I take antibiotics and antihistamines. I drink orange juice copiously. I have a fever thermometer; I keep a chart of my temperature; I keep warm, and I am usually in bed for three or four days until the fever begins to break.

When my Mexican employee starts to get the flu, he takes a concoction of herbs and keeps right on working; and he hasn't missed a day since he has been on the ranch.

Moreover, since Sam Hicks has become interested in herbs, he is doing virtually the same thing.

One thing is certain — there are medicinal properties in plants and herbs, most of which are unknown to the average modern individual.

And, apart from any medicinal value, the authentic compilation of knowledge which has been kept secret by Indians and Mexicans is a valuable contribution to our intellectual heritage.

I think this is an important book.

ERLE STANLEY GARDNER

February 7, 1966

Publisher's Preface

The medicinal uses of the plants described in this book are not based on scientific research and are not endorsed as cures by the publisher.

Acknowledgments

So many persons have contributed to the information set forth in this book that it would be impossible for me to name them all here. But to all my friends who have helped, and especially to Erle Stanley Gardner, David Hurtado, W. D. Smithers, Angel Lopez, Juanita Nejo, Inéz Romero, and Anthony Ashman, my sincere and heartfelt thanks.

S.H.

Contents

List of Illustrations		xvii
Introduction		xix
1	The Tradition of Herb People	1
2	Disinfectant and Healing Plants and Materials	5
	Yerba del Manzo, or Swamproot	5
	Melón	6
	Mescaha, or Sagebrush	8
	Yerba del Pasmo	9
3	Medicinal Herbs of the Sierra	11
	Matarique	13
	Chuchupati	14
	Yerba de la Vibora	14
	Yerba Colorada	14
	Escorcionera	14
	Pionía	15
	Bavisa	15
	Totolmeca	15
	Manzanilla	15
	Oregano	15
	Yerba del Indio	16
4	Useful Cactus Plants	17
	Charamatraca	17
	Prickly Pear Cactus, or *Nopal*	18

	Cardón	18
	Viznaga	21
	Musaro or *Garambullo*	23
5	Medicinal, Edible, and Other Useful Plants	25
	Salvia, or White Sage	25
	Chía	26
	Manrubio Blanco	28
	Estafiate	28
	Ruda, or Rue	28
	Kuanaya	29
	Ocotillo	30
	Squawbush	30
	Candelilla	30
	Popotillo	31
	Flor de San Pedro	31
	Mistletoe	31
	Ejotón	32
6	Beverage Teas	33
	Damiana	33
	Sycamore Bark, or *Cáscara de Aliso,* Tea	34
	Té de la Sierra	35
	Té del Campo	35
7	Medicines of Rural Mexico	37
8	*Alamo,* or Cottonwood, and Other Trees	63
	Eucalyptus	63
	Guatamote	65
	Elderberry, or *Saúco*	66
	Gobernadora	67
	Higuera Cimarrona, the Wild Fig Tree of Baja	68
	Dipúa	70
	Palo Fierro, or Desert Ironwood	70
	Alamo, or Cottonwood	72

List of Illustrations

PICTURE SECTION I Between pages 8 and 9

1. Angel Lopez with curative ingredients
2. David Hurtado gathering sycamore bark
3. Juanita Nejo binding basket
4. Herb doctor Luis Sui Qui
5. Uprooted *yerba del manzo*
6. Live oak bark
7. Ground *yerba del manzo*, live oak bark, and deer horn
8. *Melón* plant
9. *Melón* leaves
10. Tips of *mescaha* (sagebrush)
11. Gathering *yerba del pasmo*
12. *Yerba del pasmo* plant
13. Mexican hand-carved stone *olla*
14. *Matarique* plant and roots
15. *Chuchupati*

PICTURE SECTION II Bewteen pages 16 and 17

16. *Yerba de la vibora*
17. *Yerba colorada*
18. *Escorcionera*
19. *Pionía*
20. *Bavisa*
21. *Totolmeca*
22. Edible *manzanilla* buds
23. *Manzanilla* plant
24. Oregano
25. *Yerba del indio*
26. *Charamatraca* stalk and root
27. Inéz Romero and *charamatraca*
28. Ripe prickly pears *(tunas)*
29. Versatile *cardón* skeleton
30. Sweet, juicy *cardón* fruit
31. Towering *cardón*

PICTURE SECTION III Between pages 24 and 25

32. *Viznaga* growing
33. *Musaro*, or *garambullo*
34. Stalk of *musaro*
35. Five- and six-sided *musaro*
36. Tijuana herb vendor slicing *musaro*
37. Slicing *viznaga* after initial cooking
38. Viznaga cactus candy, or *cubiertos de viznaga*
39. Adobes fired with *cardón* and roof tiles melted by the heat
40. *Chía* plants and edible seeds
41. *Salvia*, or white sage
42. Miniature *chía* and tiny seeds
43. *Manrubio blanco*

xvii

PICTURE SECTION IV   Between pages 32 and 33

44. *Estafiate*
45. *Ruda* plant
46. *Ruda* branches and blossoms
47. Gathering *kuanaya*
48. Frances Powvall and basketwork
49. *Kuanaya*, coiled and tied
50. Hat made from *kuanaya* and pine needles
51. Splitting *kuanaya*
52. Juanita Nejo showing the stages of her craft
53. *Ocotillo* and adobe combined as building materials
54. Tony Ashman examining squawbush shoots for basket making
55. Squawbush shoots being used in coil-binding basketwork
56. *Candelilla* growing in the desert
57. *Candelilla* stem with oozing sap
58. *Popotillo* plant
59. Jointed *popotillo* stems
60. *Flor de San Pedro*
61. Mistletoe
62. *Ejotón* branch and bean pods
63. *Ejotón* growing
64. *Damiana* of Baja California
65. Elderberry blossom
66. *Té de la Sierra* leaves
67. *Té de la Sierra* plant
68. *Té del campo* plant
69. *Té del campo* blossoms

PICTURE SECTION V   Between pages 48 and 49

70. Rural Mexican garden for flower and herb growing
71. Spider web carefully preserved for use to stop bleeding
72. Dried buck deer blood taken for stomach and heart ailments
73. Deer-nose, dried and powdered, for emetic purposes
74. Scaly red eucalyptus
75. Smooth white eucalyptus
76. Shooting *guatamote* shaft
77. *Guatamote* for medicinal tea
78. Gorgonio Fernandez showing *gobernadora* poultice
79. *Gobernadora* used for tea
80. *Higuera cimarrona* figs
81. *Higuera cimarrona* tree

PICTURE SECTION VI   Between pages 70 and 71

82. Hundred-year-old *batea* made from *higuera cimarrona*
83. Bundle of *dipúa* carried by stockmen for feeding stock
84. Boys gathering wood for the Mexican village of San Ignacio
85. Iron-like *palo fierro* wood harvested with sledgehammers, not axes
86. Inéz Romero shaping a *palo fierro* block for harpooning
87. Pack-saddle blades made of lightweight cottonwood
88. Cottonwood cavalry barracks still standing at old Fort Cady
89. Leafy cottonwoods providing shade for summer siestas

Introduction

No one knows when herbs of medicinal value were first used, and few care to even venture a guess. In all probability, certain unknown early plants which produced a feeling of well-being were recognized and ingested regularly by the primates who preceded man. After the emergence of man, in the early dawn of time, there followed thousands of centuries of gastronomical experimentation by this strange, upright being, during which time he learned to select from available foods those which were best suited for his system.

There is no proof, but there is every reason to believe that in the natural course of events, our earliest forbears discovered specific roots and barks which aided their digestion after a bountiful feast. Certain soft, green plants conditioned their stomachs for the intake of food after prolonged fasting, in much the same manner that a northern black bear just out of hibernation in early spring still eats nothing for the first few days except the grasses and wild onions which shoot up rapidly in the wake of melting snows.

In recent centuries the progress of scientific medical research and the discovery of so many reliable cures for man's various ailments by dedicated doctors, have largely supplanted, among modern people, the general need for widespread knowledge of herb medicines.

One of the few remaining places where I have had the

opportunity to observe a lingering interest in herb medicines and similarly useful plants, is the desert regions of the southwestern United States and northern Mexico. Here, in spite of our space-age advancement and miraculous medicines, the native people continue to display a proud knowledge of their flora and its many practical uses.

1

The Tradition of Herb People

Regular medical services and ready availability of needed supplies are still nonexistent in so many of the remote and sparsely populated areas of Mexico that the natives, particularly the outdoor people such as ranchers, miners, and fishermen of lonely coastal waters, early inherit a superb education in the use of herbs and plants; then they proceed to expand upon that knowledge throughout their lifetime.

I know nothing of the science of botany and very little about herbs. It is the people — those who display a fascination for the flora which surrounds them and who possess a keen knowledge of the medicinal, structural, or nutritional values of this plant life — who command my attention and respect. My interest in this subject, therefore, stems not at all from the desire to become an authority on herbs and shrubs and vines, but mainly from the enjoyment of recalling the pleasant memories I hold of nightly visits around late campfires with my friends of Mexican and Indian descent. Now, almost every useful plant I recognize in the Southwest serves to remind me of a certain instance when Inéz, or Juan, or Lorenzo, or José gave me a campfire lecture on its particular virtues.

The manner in which certain herbs and plants are invariably used identically as remedies for sickness, as tonics, or for other practical purposes, regardless of distances or different languages involved, has to me become the object of fascinated reflection. In the many Indian dialects of southern California, for example, a single herb may have half a dozen different names, as well as one in Spanish and two or three in English; yet always it is used in an identical manner and for identical reasons.

David (pronounced Dah-veed') Hurtado was born and raised in the little mountain village of Yecora, Sonora, Mexico, a former sawmill town situated at the five-thousand-foot level in the timber belt of the central Sierra Madre. Like nearly all of his countrymen from the small pueblos of Mexico, David is a man of good character and of many accomplishments. He is a fine stockman and can do any kind of work entailing the use of burros, mules, horses, or cattle. He is a skillful truck driver and of sheer necessity a good mechanic. He lived many years with the Pima Indians of the Sierra Madre and learned to speak their language. He is an outstanding hunter and tracker and knows mining, timbering, and a lot about the sawmill business. His knowledge of herbs and useful plans is encyclopedic. *(See photograph no. 2.)*

Angel Lopez, formerly of Ixclan, Nyarit, is also a man of self-reliance and unusal capabilities. He attributes the cure of his stomach ulcers, shortly after he came to the United States, to the herb teas he took while he was working as a section hand on the railroad. He now has a few head of milk cows, some poultry, and a good many beehives from which he derives his principal income. He is the gentlest person with livestock I believe I have ever seen and is so considerate of his bees that his actions seem to border on the ridiculous.

    I have watched him hunting about his place on cold, spring evenings, carefully gathering up those bees too chilled and too heavily laden with pollen to fly. He gently puts them in his old felt hat, and after searching until he's sure that none will be left out to suffer further from the cold, he carries them to his car and closes them in for the night. He is an enthusiastic student of natural things and a man of infinite patience. I once visited with him as he dug a colony of ants out of the ground near his house, put them all in a fruit jar, and transplanted them several miles away, rather than exterminate them.

    Every day throughout the summer, Angel drives a couple of miles through the foothills of the Pechanga Reservation to work with his bees. Upon his return in the evening he nearly always has in the back of his car a collection of herbs that are useful as aromatic teas, stomach tonics, or disinfectants. *(See photograph no. 1.)*

    Juanita Nejo, a young Indian woman now in her middle eighties, makes annual trips to the Inaja Reservation near Julian, California, to gather herbs for her health and pine needles for the baskets she still weaves and sells. She also makes an occasional trip to the Warner Springs area for two kinds of herbs she calls in her Indian dialect *melón* and *mescaha*, both of which have amazing curative powers that I have had the opportunity to observe. *(See photograph no. 3.)*

    Herb doctor Luis Sui Qui came to Mexico from Canton, China, in 1885 when he was seventeen years old, and settled in Guadalajara. While living there he studied the use of herbs under the tutorship of other Cantonese. During the

Mexican Revolution of 1914, he made his way up the west coast to Guaymas, crossed over to Mulege, Baja California, and there for approximately forty years administered aid to the people of this beautiful oasis city through the use of both Chinese and Mexican herbs.

Presently living in the tiny village of San José Magdalena, Doctor Luis Sui Qui, age ninety-seven, still travels fourteen miles over a rocky road into Mulege twice each week to treat his patients' illnesses with the strange collection of herbs which have made him famous throughout Baja California. *(See photograph no. 4.)*

Goat rancher and fisherman Manuel Diaz, of Bahía de la Concepción, crumbles dried bark of the copal tree into the open wounds of both persons and livestock to induce quick healing. Sylvestre Arce, from the mountains north of San Ignacio, prefers to cut a limb from the soft and leafy lumboy tree and apply the ample, sticky juice to fresh cuts, blisters, and bruises.

Here is but a sampling of the kind of people who have sharpened my interest in the usefulness of plants, shrubs, roots, and the flowers of esthetic beauty and fragrance which abound throughout the arid Southwest. These people are not health faddists. They use these plants in their daily lives just as they use tortillas and beans. They have nothing to sell, no axes to grind; nor are they trying to impress anyone with their vast knowledge of nature's endowments.

From these people, and a host of other cowboys, fishermen, guides, and prospectors from the desert region of Mexico, I have gathered the following photographs and information on the useful plants of the Southwest.

2

Disinfectant and Healing Plants and Materials

Among the most important of traditional lore are those remedies for healing various sores and pains and for disinfecting. For these purposes, various parts of plants and other materials are used. Many of these preparations are equally effective for men and animals.

Yerba del Manzo, or Swamproot

The herb known as *yerba del manzo* (pronounced yerr-bah del mahn'-so), or swamproot, is found throughout the southwestern United States and northern Mexico. As its name implies, it grows in moist to swampy ground and is easily identified by its green, oval-shaped leaves and its seed stalk which shoots upward directly from the center of the plant. It grows among the tough roots of swamp grass and is easily found but difficult to dig out.

That part of the plant above ground is generally not used. The strength lies in the roots, which are frequently cooked into tea for stomach upsets. It is extremely bitter, and one small segment of root, approximately a quarter of an inch in diameter and two inches long, or that equivalent, is sufficient for one quart of tea. It should be brought to a boil and then allowed to steep. The tea is also recommended as a medicine for common colds.

Finely ground root is taken with olive oil for chronic stomach ailments. A strong, boiled solution is used as a disinfectant for bathing open wounds and swellings. Powdered root is sprinkled in open wounds. A strong, black, healing powder is made from the following ingredients:

> Bark from a living live oak tree
> Section of deer horn thoroughly burned in a wood fire
> *Yerba del manzo,* or swamproot
> Deer manure pellets

All ingredients are ground into a fine powder on a *metate* or in a hand-operated corn mill. Equal portions of powdered *yerba del manzo,* powdered deer horn, and powdered deer manure are used. A half portion of powdered live oak bark is then added. (The bark from the live oak can be used by itself, also, in making a strong disinfectant tea for washing wounds.) Powders are mixed together and dusted on surfaces of open wounds. It is used extensively for the sore shoulders and sore backs of horses and mules. (See *photographs nos. 5, 6, 7.*)

Melón

Melón (pronounced may-lone′) is widely used throughout the southwestern United States and Mexico as a powerful disinfectant and a moist healing agent.

In late August and early September, 1963, I watched the treatment of a tremendous chronic ulcer on a man's leg. The ulcer was bathed in *melón* tea (made by cooking the entire plant) twice daily, and each treatment lasted at least an hour. After the first week, signs of healing were very apparent, and the treatment was reduced to one a day.

From my limited knowledge of the situation, it appeared that the original cause of the ulcer — which was the size of a man's hand, an inch and a half deep with three inches of the shin bone exposed — could be attributed chiefly to poor circulation. After a regular cleansing and soaking with strong *melón* tea for a month, the gaping cavity filled with a healthy, pink flesh, and the man regained the use of his leg, even though the lesion had no covering of skin.

Melón is pale green and can be recognized by the fine thistledown which lines the stalks and the top and bottom surfaces of its rounded leaves. The plants generally grow in clumps of approximately six to eight inches in height and diameter. The furry thistledown covering the entire plant is soft to the touch, and there is enough of it on the leaves to make them feel almost like felt.

Teas cooked from medicinal herbs of the Southwest such as *melón* have long been used to sterilize and heal serious wounds. A long-time friend of my family, Harv Nibarger, had once been shot in the knee by an unknown assailant using a high-powered rifle, while Harv was driving a bunch of cattle. He rode as fast as he could to an Indian camp in the area where one of the women helped him into her brush hogan and administered first aid. The Indian woman splinted the shattered knee joint and then rigged a tripod over Harv's leg from which she suspended a pot of herb tea. The pot was carefully arranged so the tea would slowly drip into the wound.

I've forgotten how long Nibarger told us it had taken for the wounded knee to heal, but I'll never forget the expressions of gratitude he lavished on the Indian woman he believed had saved his leg. He used to sleep in the ranch bunkhouse with my brothers and me, and at bedtime when he undressed we would crane our necks to look at the rigid, grotesque limb Harv was always so proud to have kept.

During the Mexican Revolution in 1916, Pancho Villa received a similar knee wound and underwent the same dripping herb-tea treatment in a nondescript little ranch house near Casas Grandes in the state of Chihuahua and there escaped detection until he had sufficiently recovered to ride again. He was more fortunate than Nibarger in that, according to history, his knee healed up without being stiff. (*See photographs nos. 8, 9.*)

Mescaha, or Sagebrush

Mescaha (pronounced mes-ka′-ha), one of the most prevalent aromatic shrubs of the Southwest, is commonly used as a medicinal or disinfectant tea. This tea is bitter and unpalatable if cooked too strong. As an effective antiseptic for bathing wounds, the brush tips and leaves are vigorously boiled until the tea is deep green. It is especially good for washing and healing wire cuts on horses. Several years ago a weak tea was customarily taken in the spring of the year as a tonic by ranching families of the West, and the frontier women of the Great Plains states used sagebrush tea regularly as a hair rinse.

Mescaha, or sagebrush, grows at higher elevations, from two thousand to ten thousand feet, and its wood burns hot. It has always been a popular fuel for branding-iron fires. An identifying factor is the snubbed-off end of each leaf which appears to have been cut with pinking shears, leaving three rounded points. Slate gray, the tips of the brush and leaves are always cooked together and, when boiled, turn green.

At higher elevations where the brush grows rank, its tips and leaves are broken off by the handful and bound with

1. Angel Lopez with curative ingredients
2. David Hurtado gathering sycamore bark
3. Juanita Nejo binding basket
4. Herb doctor Luis Sui Qui

5. Uprooted *yerba del manzo*
6. Live oak bark
7. Ground *yerba del manzo*, live oak bark, and deer horn

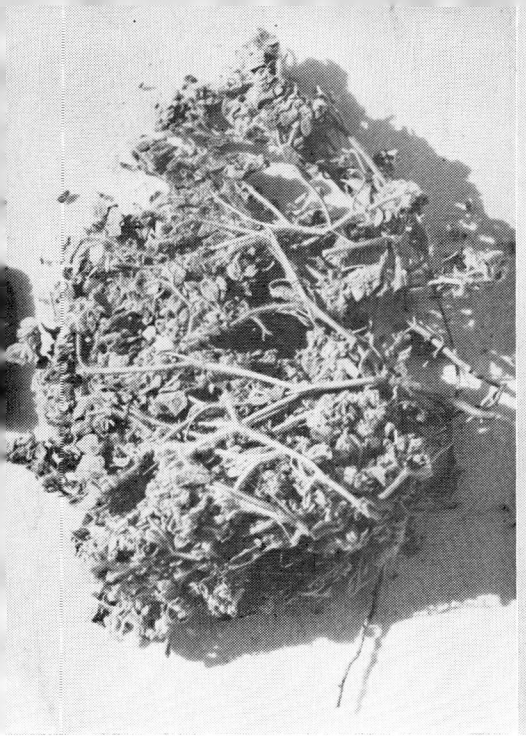
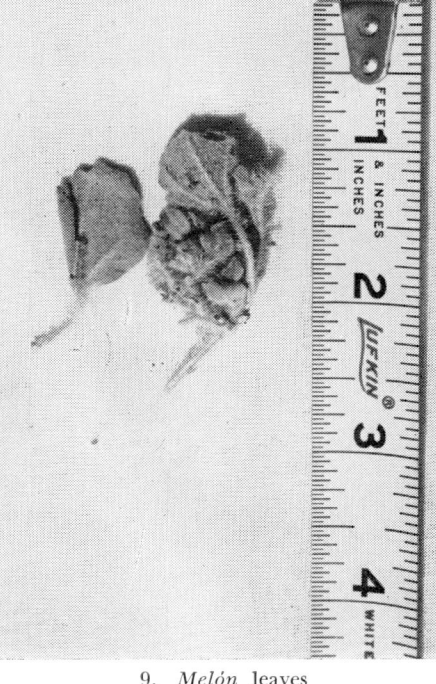

8. *Melón* plant

9. *Melón* leaves

10. Tips of *mescaha* (sagebrush)

11. Gathering *yerba del pasmo*

12. *Yerba del pasmo* plant    13. Mexican hand-carved stone *olla*

14. *Matarique* plant and roots   15. *Chuchupati*

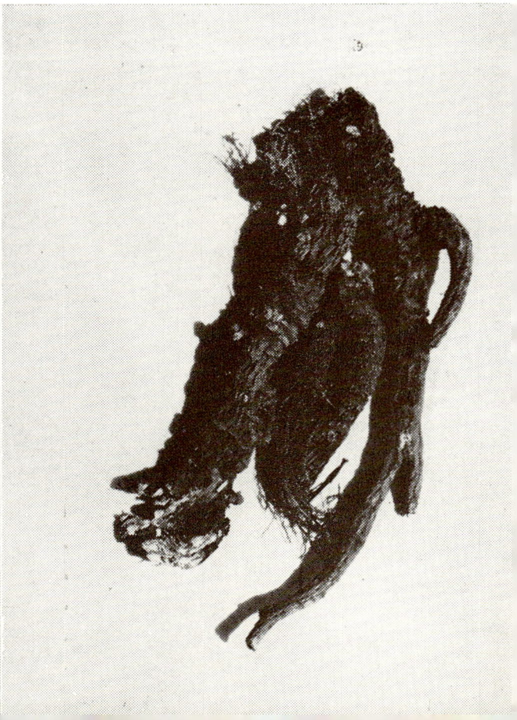

string to preformed, circular wire coat hangers to make thick, silver-colored aromatic wreaths. (*See photograph no. 10.*)

Yerba del Pasmo

Yerba del pasmo (yerr'-bah del pahs'-mo) is a low-growing herb with healing properties. Tea cooked from *yerba del pasmo* is used to wash infected wounds and is taken orally simultaneously, as it is believed that drinking the tea promotes healing. Once Tony Ashman, of Pechanga, treated the infected feet of a factory worker with *yerba del pasmo*. The worker had been told to seek other employment because of the condition of his feet. Following the treatment, his foot condition disappeared. Slow-healing ulcers are also treated with *yerba del pasmo* tea to which salt has been added. (*See photographs no. 11, 12.*)

3

Medicinal Herbs of the Sierra

These popular medicinal plants grow in the mountain regions of northern and central Mexico at elevations of approximately four to six thousand feet. Their standard uses are well known by all the people living on ranches and in the mountain pueblos of the Sierra Madre.

Of the many beneficial native plants depended upon by the people living in isolated, rugged areas of Mexico, the plants shown here are by far the most widely used. Their names are common household words; and at least some, if not all, of these dry medicinal roots, or leaves, or blossoms, can be found stored away in fruit jars in the kitchen of almost every mountain dwelling. All of the herbs mentioned here have their medicinal qualities contained in the roots, with the exception of *manzanilla* and oregano. The useful portions of these two plants grow above ground. All the plants seek partial shade and grow rapidly after seasonal rains.

In the United States oregano is thought of almost entirely as a spice. In Mexico it is cherished as a curative plant of many uses. Oregano tea is drunk as a remedy for coughs and colds. It is used as a disinfectant and healing agent for burns, pimples and skin irritations, and infected wounds. In the home women add liberal quantities of the flavorful leaves to their chili sauces and other spicy dishes, while their *ranchero* husbands mix finely ground oregano leaves with lard or tallow and use the salve to heal wounds on livestock.

Almost every baby in Northern Mexico has been fed *manzanilla* tea by its mother. It has long been the standard home treatment for tiny babies having difficulties digesting their milk. As they grow older, Mexican babies are given bottles of *manzanilla* tea between regular feedings. *Manzanilla* buds, or blossoms, have always been a popular item in Mexican drugstores, and they can be picked almost anywhere once a year while the ground is moist. The plant's round, yellow buttons always appear ready to burst into flower, but they never do. Picked and eaten fresh, the mintlike flavor of the buds is delicious, especially so after meals.

When you live in a country where abundant groceries are always readily available, and tempting varieties of food and drink constantly flash before the public's eyes through our modern media of advertising, it is hard to believe that anyone, anywhere, would ever have the need to stimulate his appetite. Unfortunately, this situation does not exist in many countries. In those places where the daily diet of people remains the same for as many as three hundred and fifty days of every year, an herb that will stimulate appetites and make the same rough food, which has been served for at least a thousand consecutive meals, look and smell delicious — this is indeed a valuable medicine. Therefore, when certain people tell you of an herb which is a general stomach tonic and appetite stimulator, that information is of great importance to them, and it should not be taken lightly by those fortunate enough to have a variety of delicacies readily available to them at all times.

Matarique

Matarique (pronounced mah-tay-re'-kay) grows only in the shady seclusion of the Sierra Madre Mountains of northern Mexico and only during the rainy summer months when the high ground on which it thrives is wet with moisture. Its green foliage resembles endive, thus making is especially easy to identify. The curative powers which are attributed to the plant are contained in an ample, intertwined root system which, after being pulled from the soft earth, readily releases its pleasant aroma and a spicy, pepsin taste. After chewing briefly on a segment of the root, a person feels as though he has just brushed his teeth, then gargled with a pleasing mouthwash. During hot weather *matarique* roots are used in early every stone water *olla* in the *ranchos* and *pueblitos* of Sonora and Chihuahua, Mexico.

The roots of a single plant may remain in a water *olla* for as long as two months at a time, freshening new water as it is added and providing a stomach-soothing drink in an area where, of necessity, water must be consumed in volume. Water drunk from an *olla* containing *matarique* roots rapidly quenches a person's thirst and causes him to perspire less.

Tea cooked from the roots is taken for the relief of severe back pains and as an unfailing remedy in the treatment of jaundice. Besides being considered an ideal blood tonic, it is given to babies for colic and taken by adults as a diuretic.

Its roots, when ground on a *metate,* are used as a poultice to heal open cuts, and a strong solution cooked from the roots is used as an antiseptic wash in treating slow-healing wounds. (*See photographs no. 13, 14.*)

Chuchupati

The root of *chuchupati* (pronounced choo-choo-pah'-te) is crushed on a *metate,* then placed directly on the area affected by scorpion or bee stings or spider bites. If the victim is particularly allergic, a tea cooked from *chuchupati* is drunk which is believed to reduce swelling in the throat resulting from insect stings. The tea, very bitter, is also taken for biliousness. *(See photograph no. 15.)*

Yerba de la Vibora

Yerba de la Vibora (pronounced yerr'bah day lah vee-bo-rah) is Mexico's standard rattlesnake bite remedy. The root is crushed on a *metate,* or sometimes cooked, then placed on the snake bite after it has been bled and sucked. The crushed herb is cooked to make a pleasant tea which tastes very much like *matarique.* Snake-bite victims are urged by native herbists to drink this tea in quantity. *(See photograph no. 16.)*

Yerba Colorada

The *yerba colorada* (pronounced yerr'-bah co-lo-rah-dah) is crushed and cooked into a tea which is taken for chest and back pains and for flu. The tea is very red and has a pleasant taste. The crushed root is cooked with sugar to make a popular cough syrup. *(See photograph no. 17.)*

Escorcionera

Tea from the crushed root of *escorcionera* (pronounced es-cor-cee-oh-nay'-rah) is drunk in large, regular doses by persons having stomach ulcers. Poultices of the crushed root are used to heal open sores and to reduce swelling and discoloration from bruises. *(See photograph no. 18.)*

Pionía

After *pionía* (pronounced pe-o-nee′-ah) is crushed on a *metate* and cooked, the tea is taken to stimulate appetites. It also is taken to stop stomach pains and is believed to be especially good for children's stomachaches. The root, finely ground and then toasted, is taken with water to stop diarrhea. Crushed root is packed around aching teeth to stop pain. *(See photograph no. 19.)*

Bavisa

Bavisa (pronounced bah-vee′-sah) is a general stomach tonic and is used exactly as *pionía* except *bavisa* is not taken for diarrhea. *(See photograph no. 20.)*

Totolmeca

The root of *totolmeca* (pronounced to-tol-may′-cah) is broken into small pieces and cooked. The tea is taken principally for female disorders. It is also considered an excellent kidney medicine and is taken to relieve severe back pains. It is carefully avoided by pregnant women. *(See photograph no. 21.)*

Manzanilla

The round to conical-shaped buds of *manzanilla* (pronounced man-sah-nee′-yah) are brewed into a popular tea which is taken for upset stomachs. It is used almost universally by the mothers of small babies throughout northern Mexico as a stomach tonic and is fed to children of all ages from either a spoon or a bottle. The tea is also used for washing cuts and abrasions. *(See photograph no. 22, 23.)*

Oregano

The leaves and small stems of oregano are cooked into

a tea which is taken for coughs and colds and used as a disinfectant for washing wounds, burns, and skin irritations. Its leaves are used as a spice in cooking, and when finely ground they are mixed into a healing salve. *(See photograph no. 24.)*

Yerba del Indio

Easily recognized by the shape of its arrow-point leaves, the roots of the *yerba del indio* (pronounced yerr'-bah del een'-de-o) plants are much sought after as a general tonic. It can be found growing in many types of mountain terrain but seems to thrive principally on the dry floors of rocky arroyos. The freshly dug, carrot-shaped roots are good to eat, and they stimulate the appetite. Tea cooked from the roots is also pleasant tasting and is taken to soothe upset stomachs and aid digestion. *(See photograph no. 25.)*

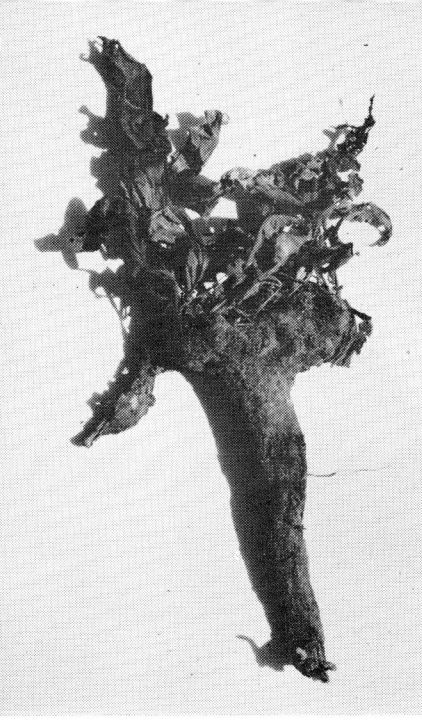

16. *Yerba de la víbora*

17. *Yerba colorada*

18. *Escorcionera*

19. *Pionía*

20. *Bavisa*

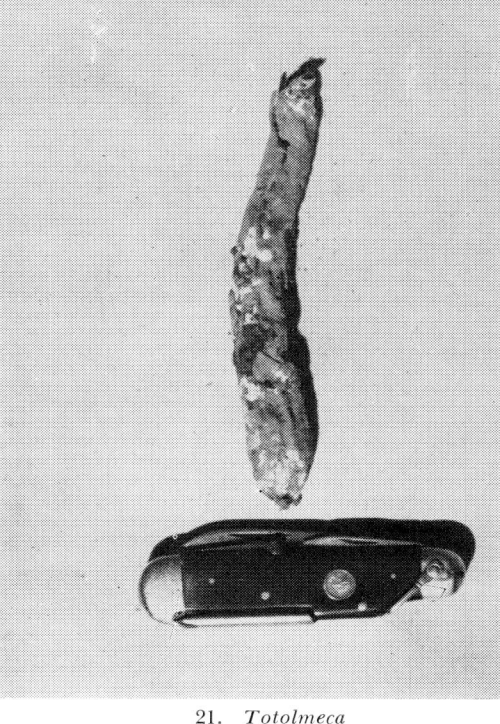

21. *Totolmeca*

22. Edible *manzanilla* buds

23. *Manzanilla* plant

24. Oregano
25. *Yerba del indio*

26. *Charamatraca* stalk and root
27. Inéz Romero and *charamatraca*

28. Ripe prickly pears (*tunas*)

29. Versatile *cardón* skeleton

30. Sweet, juicy *cardón* fruit

31. Towering *cardón*

4

Useful Cactus Plants

Various cacti are indigenous to southwestern United States and to northwestern Mexico. Certain of these plants are useful in medicinal preparations, even being cultivated for this purpose. Others are edible and can supply food in arid areas for both men and animals.

Charamatraca

Charamatraca (pronounced chah-rah-mah-trah'-cah), a slender cactus, has all the attractive, uniform markings of a snake and is inclined to climb other plants like a vine. In Baja California it is found only in the area between Bahía de los Angeles and El Arco. Each willowy stalk of *charamatraca* is nourished by an enormous root. The root is soft and pliable, almost transparent, and is comparable to a plastic bag filled with a yellow liquid. It lies just beneath the surface in loose, sandy soil and can be easily dug out by hand. *Charamatraca* has spines, but they are so small that they can hardly be seen.

Whenever a native Baja Californian of this central region is suffering severe back pains, he gathers one of these watery roots, strips the juice from it into a cloth, then places the cloth on his back until the pain subsides. (*See photographs no. 26, 27.*)

Prickly Pear Cactus, or *Nopal*

Prickly pear cactus, or *nopal* (pronounced no-pahl′), is much utilized as a food in desert regions. The broad, flat leaves of this plant can be roasted on the coals of an open fire and eaten or can be peeled and cubed to make the popular Mexican dish called *nopalitos*. The fruit, called *las tunas* by Spanish-speaking people, is easily peeled and, though full of small, hard seeds, is delicious when ripe. Its delicate flavor is probably appreciated the most after the fruit has been made into jam or jelly.

In the past this cactus has been widely used as livestock feed. During periods of severe drought ranchers used to burn the spines off the leaves with pressurized gasoline torches. Their cattle, upon hearing the roaring sound of the generating gas burners, would come running to eat the freshly singed cactus. When feed is unusually scarce and cattle are desperately hungry, they will eat several kinds of cactus besides prickly pear, spines and all, and survive. (*See photograph no. 28.*)

Cardón

We were making a road by hand for our four-wheel-drive equipment across a rough arroyo in central Baja California, called El Alambrado when our friend, Inéz Romero, a Yaqui Indian, seriously mashed the fingers of one hand between two giant boulders. While the other members of the party continued rolling the bigger rocks out of our proposed line of travel, I extricated a dusty first-aid kit from the varied assortment of camp supplies and equipment that we regularly carry on an Erle Stanley Gardner ground expedition into Baja.

Inéz's index finger was badly split, and his middle finger had been flattened with such force that it appeared as though an inner explosion had forced the raw flesh out through the ragged tears along both sides of the finger just below the nail.

Inéz Romero is a colorful Baja Californian who does everything with gusto, whether it is harpooning turtles, hunting mountain sheep, or getting the FWD pickup he drove for us stuck in loose sand. So I knew instinctively before I examined his hand that he had smashed it with characteristic exuberance.

While I pondered the ample contents of the first-aid kit — it was the size of a suitcase — wondering just which of all the various items would do Inéz's fingers the most good, he stopped jumping up and down and blowing on his fingers long enough to tell me that all we needed from the box of fancy medical and surgical dressings was some gauze for bandaging. After the pain had subsided a little more, Inéz explained that all he wanted on his injured fingers were some slabs of *cardón* (pronounced car-done').

I had a fleeting impulse to explain the super disinfectant qualities of merthiolate to him and argue that it was surely better than a couple pieces of cactus meat hacked out of a *cardón* by an unsterilized belt knife. But on a quick second thought, I shut my mouth and dutifully followed Inéz to the nearest *cardón*. Under his direction I peeled the spines and green bark away from a section of its fluted surface and cut out a hand-sized rectangle of the tough, white pulp. Inéz had me slice two thin wafers from the rectangle, wrap them around his mashed fingers, then bandage them with gauze and tape.

In camp that night he admitted that his fingers hurt a little, but by the next morning he insisted that all the pain had subsided, and he continued to drive the pickup with his usual flair. After supper the following evening Inéz reluctantly allowed us to remove the bandages, claiming there was no need to do so for another two or three days, and his fingers emerged so startingly white that they resembled a couple of dead fish. Inéz nodded approvingly at his mutilated *dedos* and quietly remarked that was the way they should look.

He asked us to please wrap them up again with the same *cardón* and bandages. We compromised by cutting some fresh *cardón* wafers and wrapping his fingers with new gauze and tape — but this time it was black electrician's tape because the moisture from the *cardón* wafers kept loosening the other kind.

In a week, with no further attention, Inéz's fingers were completely well and, according to him, at no time during the healing process had they been sore. He was disgusted with us for being concerned and explained that he believed the pulp of a *cardón* not only has a built-in painkiller but also contains a disinfectant and a powerful healing agent as well.

The dry wood from *cardón* skeletons burns terrifically hot. At Mulege all of the baked adobe bricks and roof tiles used in the modern Club Aéreo Hotel were fired with wood from dead *cardones*. The site on Don José Gorosave's Llanos de San Ignacio Rancho, where the adobes and roof tiles were made, is generously littered with stacks of crisp clay tiles that became so hot in the firing process they melted and ran together like glass in a furnace. When tapped with a stick, these misshapen clusters ring like bells of differing tones.

The hard ribs from the skeletons of *cardones* can be found on nearly every *rancho* in Baja where they are either tied together with rawhide or wire, or nailed to make corrals, yard fences, and the walls of buildings.

The ripe *cardón* fruit is eaten with relish by both people and livestock. Hungry cattle rub against the trunks of the huge cactus to dislodge the fruit, then eat it as soon as it hits the ground. A curious thing about *cardones* is the fact that the drier the seasons become in Baja, the more plentiful and robust grows the fruit.

Smart campers in Baja California study the condition of nearby *cardones* before rolling out their beds at night. Sometimes a heavy limb or, on occasion, an entire massive *cardón* measuring fifty feet and weighing tons, will come crashing down in a windstorm with a terrifying thud — the memory of which spoils the campers' sleep for several nights to come.

With the exception of a small area in western Sonora, *cardones* grow only in Baja California. Beginning at an imaginary line approximately two hundred miles south of the international boundary, they dot the remainder of the peninsula and are found growing in profusion in many inland areas similar to the familiar thick stands of *cardones* which thrive along the eastern and southern reaches of Bahía de Concepción. (*See photographs no. 29, 30, 31, 39.*)

Viznaga

Viznaga (pronounced vez-nah'-gah), or purple hedgehog cactus, can be eaten, but it is more a novelty or a survival food than it is gourmet's delight. It is filling and rather tasty if it is first roasted in a bed of coals for an hour or so until the fiber of its white meat is broken down. It is peeled, sliced in quarter-inch thick slabs across the grain, then breaded and fried the same as eggplant. Its nutritional value is debatable.

Cubiertos de viznaga, Spanish for cactus candy, is made by first soaking large, rectangular wafers of the *viznaga* cactus in a strong lime and water solution for approximately one half-hour. The wafers are then slowly drained on boards or palm fronds placed outside in the night air during the time of year when heavy dew is a certainty. If the wafers are drained under almost any other condition, they harden slightly on the outside, trapping lime water inside the slabs and rendering them useless for human consumption.

Standard measurements of the ingredients for a batch of cactus candy are ten pounds of soaked *viznaga* slabs, twenty pounds of *panocha* (brown sugar), and five gallons of water. The water and sugar are brought to a boil in a large pot over an outside fire, then the slabs are added and cooked until they become candied, which takes about twenty-five to thirty minutes. They are then lifted from the boiling pot with a wire mesh scoop and spread out on a table to cool.

Conserva always follows in the wake of a batch of *cubiertos,* and it has long been a popular Mexican dessert. Here again *viznaga* is used, but for *conserva* it is cut into cubes or chunks of one half-inch and smaller. This chopped *viznaga* meat is also soaked in lime water and drained in the same fashion as *cubiertos.* After the *cubiertos* are dipped out of the boiling sugar syrup, about ten pounds of the cubed *viznaga* is poured into the same pot. This batch is cooked for another twenty-five or thirty minutes, but care must now be taken to prevent overcooking. Prolonged boiling will cause the syrup to harden after it cools, thus changing the taste and spoiling the *conserva* for all practical purposes. It should be cooked just long enough that, after cooling, the syrup is still liquid with the candied *viznaga* bits remaining in a state of suspension. While it is still warm, it is poured into five-gallon honey cans for storage, in which it will keep for years.

Conserva is customarily served at room temperature in a bowl or cup and eaten with a spoon. It is sold daily in the stores of small villages and cities throughout Mexico and can be purchased in any quantity from a small helping in a bowl held in the grimy hands of a child to a five-gallon canful, which weighs about sixty pounds.

Due to the efficiency of modern transportation, *viznaga* and other varieties of desert cacti are rarely burned and used for cattle feed any more in the Southwest. Isolated instances of cactus-burning still prevail where occasionally an emergency exists, but the general practice is no longer widespread. In view of this fact, it would seem that we should enjoy a noticeable increase in the cactus plants of our deserts. Unfortunately, this is not the case.

Certain regions of the southwestern United States have suffered so long from acute shortages of rain that many of our most picturesque and interesting species of cacti no longer grow abundantly. The colorful *viznaga* is one of these species which, as of the present, is definitely not on the increase. Let's save them for photographic pleasures and emergency rations only. (*See photographs no. 32, 37, 38.*)

Musaro, or Garambullo

From the sparsely settled mountain regions to the east, Indian-women herb vendors arrive daily in the city of Navajoa, Sonora, Mexico. Most of them spread out in the town market their displays of curative plants and edible herbs gathered from the canyons of the Río Mayo and its tributaries, and there they await the approach of their regular customers. Other women balance heavily laden baskets of organic medicines on their heads and proceed through the streets, selling from door to door. Among the strange variety of medicinal leaves, barks, and roots which fill the women's baskets, thick stalks of green cactus are noticeably more abundant than any other type of plant. This cactus is called *musaro* (pronounced moo-sah'-ro) on the mainland of Mexico or *garambullo* (pronounced gah-ram-boo'-yo) in Baja California. A highly concentrated tea cooked from the sliced stalks has long been a well-known medication for ulcerated stomachs. (Warm poultices of the plant's meat are also applied for temporary relief to the swollen joints of people suffering from arthritis.) These dark auburn-haired, blue-eyed Mayo women *curanderas* first introduced *musaro* to the Mexican people as a curative plant.

Musaro, or *garambullo,* tea is made by slicing fifteen or twenty cross sections about two inches in length from the stalks of cactus. These slices are then placed in a container large enough to hold five gallons of water and boiled for eight or ten hours until the liquid is reduced to approximately one gallon. People using this treatment for serious stomach ailments drink the tea in enormous quantities.

Throughout the state of Sonora, both on ranches and in villages, *musaro* plants are raised in vegetable and flower gardens for medicinal use in homes. Gathered from the desert, the cactus sells in Navajoa and Ciudad Obregón at the rate of four or five stalks, each from one to two feet in length, for a dollar.

In Tijuana, where America's high cost of living is reflected across the border, the plant presently sells for fifty cents a slice. It can be bought in some of the fruit and vegetable stores on First Street and also found at an herb vendor's stand in El Mercado Municipal, the principal market in downtown Tijuana.

Musaro, or *garambullo,* is always a bright green and it grows somewhat like *pitahaya.* Its stalks may have either three, five, or six ridges and can easily be identified by the evenly spaced, star-shaped clusters of spines which grow only on the crests of the ridges. It can be found frequently, and in many places growing in abundance, along most of the length of the Baja California peninsula. (*See photographs no. 33, 34, 35, 36.*)

32. *Viznaga* growing 33. *Musaro*, or *garambullo*

34. Stalk of *musaro* 35. Five- and six-sided *musaro*

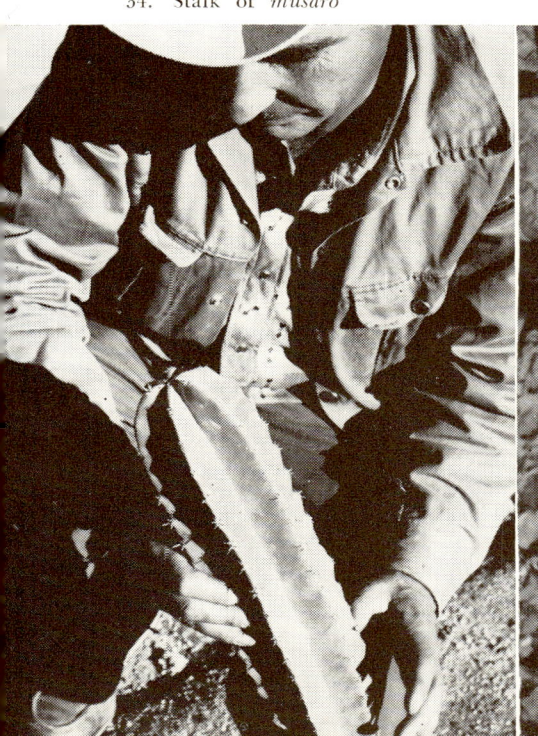

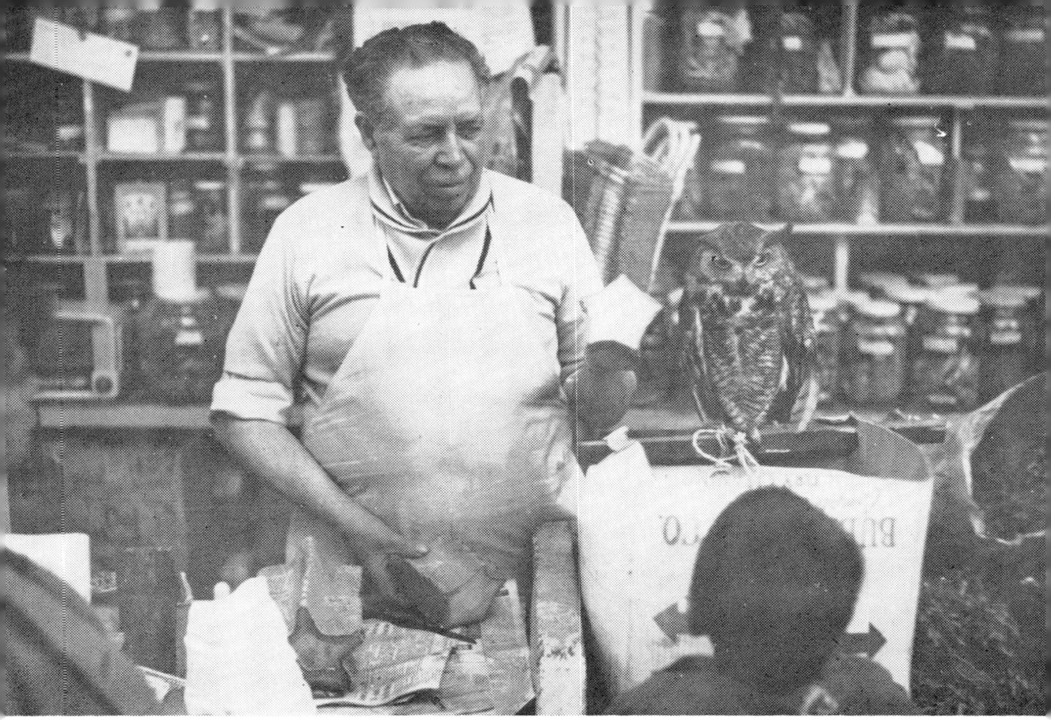

36. Tijuana herb vendor slicing *musaro*

37. Slicing *viznaga* after initial cooking

38. *Viznaga* cactus candy, or *cubiertos de viznaga*

39. Adobes fired with *cardón* and roof tiles melted by the heat

40. *Chia* plants and edible seeds
41. *Salvia,* or white sage

42. Miniature *chia* and tiny seeds
43. *Manrubio blanco*

5

Medicinal, Edible and Other Useful Plants

Certain medicinal, edible, and other useful plants have been depended upon by Indian and Mexican inhabitants of the desert area for untold years. Some of these are now becoming available commercially; others are still regarded as common pests by the untutored observer.

Salvia, or White Sage

Salvia (pronounced sahl′-ve-ah), or white sage, grows abundantly throughout southwestern United States and northwestern Mexico. Tea cooked from its leaves is taken internally as a poison-oak remedy. It is also used as an external treatment, and areas affected by poison oak are generously bathed in the same solution.

Salvia was used extensively in the past by the Indian tribes of southern California as an aid to childbirth. Quantities of the tall, white plants were burned, and the expectant mother's body was completely covered with the hot ashes. The treatment was continued until delivery began. Frequently the newborn babies, especially males, were "cooked" in the hot *salvia* ashes. These "cooked" babies are reputed to have consistently grown up to be the strongest and healthiest members of their respective tribes and are claimed to have been immune from all respiratory ailments for life.

Teas of varying strength and a tincture made from powdered, dry *salvia* leaves soaked in alcohol, are taken internally as an aid to digestion, as a remedy for excessive urination, and also as a sedative.

The *salvia* blossom is a good producer of expensive, almost water-clear honey. (*See photograph no. 41.*)

Chía

Each spiny, leafless *chía* (pronounced chee'-ah) blossom contains a prodigious quantity of seeds. In early summer, after the blossoms have dried and the seeds are ready for harvest, the plant is picked, and the blossoms, usually six or eight to each plant, are crushed by hand. The seeds, one held in the tiny tubular base of each spine, are then easily shaken into a basket. They spill out freely — almost a half-teaspoonful to the blossom. Spines and bits from the crushed pods are later winnowed out by tossing the seeds in a light breeze. The spines of the blossoms are not nearly so formidable as they appear; however, while one is crushing them, gloves are recommended for soft, uncalloused hands.

Two distinct types of *chía* grow abundantly in the sandy desert lands of the Southwest and Mexico after rainy seasons. To the layman the plants are identical in every detail except one grows in miniature.

Ancient peoples of these arid regions depended heavily

on *chia* as a staple food and, in all probability, it was harvested and eaten by them even before corn was. It is still used regularly by the descendents of the early people who first discovered its usefulness, and in just the last year *chia* has become available to the public in health-food stores of southern California.

Chia seeds from the minature plants are used by the native people more as an eye treatment than as a food. These tiny seeds can be ground on a *metate* into a meal and eaten, or soaked in liquid and drunk in the exact manner that the larger seeds from the taller, more robust plant are utilized. But, primarily because the yield is less, only a few small jars of the miniature *chia* are harvested annually. They are used in the households of local people who sprinkle the fine seeds into their eyes at night to remove foreign particles from them and to soothe irritation.

Both varieties of ripe *chia* seeds are hard and smooth so long as they are dry. When placed in water, or any liquid for that matter, they rapidly begin to soften and swell. The seeds, normally ranging from tan to black, change to a faint blue and increase their original size by four or five diameters. If plain water surrounds the *chia seeds*, it turns to a clear gelatin, and the entire tapioca-like substance is pleasant to eat, especially so if a little brown sugar is added.

As the many modes of transportation now reaching into the far-flung regions of the southwestern United States and of Mexico increase the availability of a variety of foods, it is interesting to note that *chia* is still being harvested and eaten by the people whose forefathers depended so largely upon it. Of these people I have asked the following questions: Is *chia* gathered and eaten because of its nutrition and taste? Or, possibly, is it because it is free? Or, is eating it simply a custom?

Answers vary from one family to another, but the real reason it is still in use probably lies in the combination of all three. (*See photographs no. 40, 42.*)

Manrubio Blanco

This well-known whitish-green plant flourishes in most of the Southwest and is considered a nuisance by most people because of the tenacious manner in which its cocklebur seed pods cling to clothing and to domestic animals. *Manrubio blanco* (pronounced man-roo'-be-o blahn'-co) does have medicinal properties, though, and its tea is a good remedy for biliousness. When ripe, the seeds can be easily separated from the pod clusters, and they are pleasant to eat. (*See photograph no. 43.*)

Estafiate

This plant is commonly regarded as a prolific, useless dooryard weed. However, *estafiate* (pronounced es-tah-fi-ah'-tay) tea has a soothing effect on nerves and is widely taken by pregnant women to alleviate morning sickness. A strong, cooked solution of *estafiate* tea is also given as a drench to horses in the treatment of worms. (*See photograph no. 44.*)

Ruda, or Rue

Ruda (pronounced roo'-dah) is one of the most prevalent home-remedy herbs in the Southwest. It is frequently found growing in Mexican households either as a potted plant or as a shrub in the flower garden where its yellow blossoms quickly change from flowers to thick pods which, when crushed, release an exhilarating fragrance that opens the sinus cavities much the same as menthol does. Its aromatic leaves and pods are cooked into teas of varying strengths which are taken by all for stomach disorders and by women as an aid to menstruation. *Ruda* tea is carefully avoided during pregnancy. Its leaves, slightly crushed and rolled into small plugs, are inserted into the ear to relieve earaches. (*See photographs no. 45, 46.*)

Kuanaya

In the Inaja dialect of the Mission Indians of southern California, the tall, tubular grass which, in expert hands, is split evenly into three long, uniform strands for basket-making, is called *kuanaya* (pronounced coo-ah-na'-ya). I have never heard this plant called by any other name in any other language, although I'm sure sufficient research would show that it has been scientifically classified. Basket-makers lose no time in splitting these tough stalks while they are fresh and pliable. About one-third of the top of each pointed stem is cut off and thrown away because it lacks strength. Then the square-cut end of the hollow stalk is carefully divided into three parts with the point of a knife, the end of one strand held in the teeth while the other two are pulled outward with the hands. The strands, after being separated, are next tied in neat coils and stored until needed. They are soaked briefly in water just before they are used in the ancient coil-binding process characteristic of much southwestern basketry.

Each stalk of the long grass has a dangerously hard stiletto point which seems to probe at hands, wrists, and eyes while it is being pulled from the damp earth where it grows. For a few inches at the base of each stem there is a permanent deep brown color. Higher up, the stalk is dark green, but soon after picking this color fades; and by the time the strands are ready for use, they graduate in color from yellow to tan to brown. These varying colors enable the basketmakers to weave into their art the unique designs which ornament their work. For certain designs, where a great deal of solid color is required, some basketmakers dye their *kuanaya* strands with berry juices and, in more recent years, with water colored by rusty nails and bits of iron.

This tall, fibrous grass, which grows only in the mountains near the source of fresh water, is rapidly disappearing. Due to persistent drought conditions in the Southwest and the diminishing number of mountain springs, *kuanaya* is hard to find. It is now gathered carefully and unselfishly shared by those who use it in their handicraft. *(See photographs no. 47, 48, 49, 50, 51, 52.)*

Ocotillo

The *ocotillo,* also known as coachwhip, Jacob's-staff, and vine-cactus, is found in desert areas from the western part of Texas to the southern part of California and south into Mexico. The plant usually has several slender, spiny branches and may grow as high as twenty feet. It is used as a hedge plant at times. Fences may be built of it, and it has also been used in combination with adobe mud as a building material. *(See photograph no. 53.)*

Squawbush

Each summer the squawbush sends forth long, slender shoots which are gathered and used in basketmaking. The slender shoots split easily into uniform strands two to three feet long. The bark is then peeled from the outer surfaces, leaving the strands a glossy, natural white. The fine wooden strands are soaked and used to bind the small center sections of many of the coil-binding type of baskets of the Southwest. *(See photographs no. 54, 55.)*

Candelilla

This desert plant takes its name from its use in candlemaking. The stalks of *candelilla* (can-day-ee'-yah) are swollen with a sticky, white juice. By cooking the plant, a wax used for making candles is obtained. Because of costly processing difficulties, rather than a shortage of *candelilla,* commercial attempts in Baja California to extract candle wax from this plant have not been successful. *(See photographs no. 56, 57.)*

Popotillo

Popotillo (pronounced po-po-tee'-yo) is probably the best-known desert herb used for making medicinal tea. Often called squaw tea, this leafless, green shrub grows abundantly throughout the Southwest and is taken regularly by many as a kidney medicine and as a general tonic. It is used as a laxative and for jaundice and gonorrhea. Poultices of ground *popotillo* are applied to snake bites. Most Indian and Mexican people who use *popotillo* claim that it is not medicinally valuable if the inside pith of the round, jointed shoots is not reddish-brown. It is available in nearly all health-food stores. (*See photographs no. 58, 59.*)

Flor de San Pedro

The *flor de San Pedro* (flor day sahn pay'-dro), whose name means Saint Peter's flower, undoubtedly takes its name from the trumpet shape of its blossoms. It is used chiefly as a quick, effective cure for diarrhea. The tea is cooked mostly from the flowers, though a few leaves are used also. It soothes upset stomachs and aids digestion.

During the spring and summer months, the plants bloom profusely. Pechanga Indians say that plants with red blossoms are male and plants with orange, female. Color of the blossoms seems to make no difference, however, in the plant's medicinal properties. (*See photograph no. 60.*)

Mistletoe

Mistletoe, that parasitic plant of ancient fame and reverence, is commonly found growing in clumps on many varieties of trees in the Southwest and Mexico. It is generally disliked, as it eventually destroys most of the trees it infests. In recent years, however, it has been harvested commercially in some localities and shipped to distant cities to be used in holiday decorations at Christmas.

From ancient times it has been supposed to have curative powers, and in certain areas it is used today in making medicinal poultices. Mistletoe leaves and berries are cooked with rice, and the mixture is used as a poultice to bring boils to a head and to draw pus from infected areas. *(See photograph no. 61.)*

Ejotón

The large, crusty pods of the *ejotón* (pronounced ay-ho-ton') grow profusely from the branch tips of a fiercely spined, tough desert shrub. Gray-green *ejotón* bushes are found predominantly along the banks of dry watercourses from the Bahía de los Angeles area of Baja California on south past Bahía de la Concepción. The thick, unyielding brush ranges from approximately four to seven feet in height and because of its spines is impenetrable. Both the green and the blackened, dry pods can be found on the same shrub. The black, dry pods are ground on a *metate;* tea is then cooked and taken as a treatment for rattlesnake bites. The beams which develop and drop from the opened pods are bitter and have no medicinal or nutritional value. *(See photographs no. 62, 63.)*

44. *Estafiate*  45. *Ruda* plant

46. *Ruda* branches and blossoms 47. Gathering *kuanaya*

48. Frances Powvall and basketwork

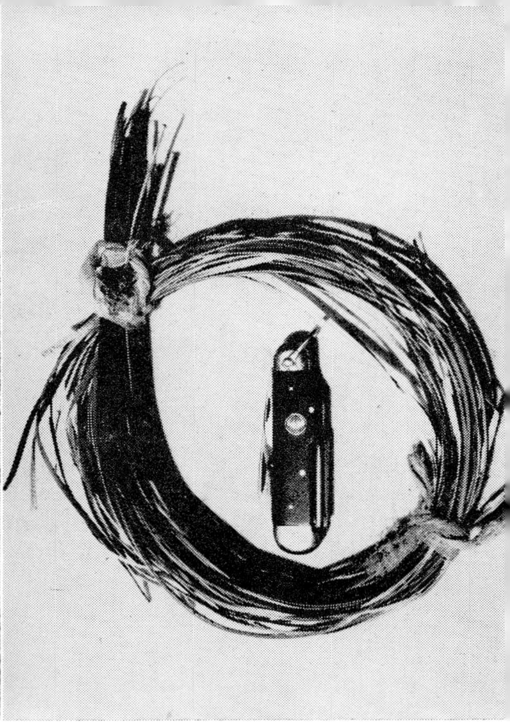

49. *Kuanaya*, coiled and tied

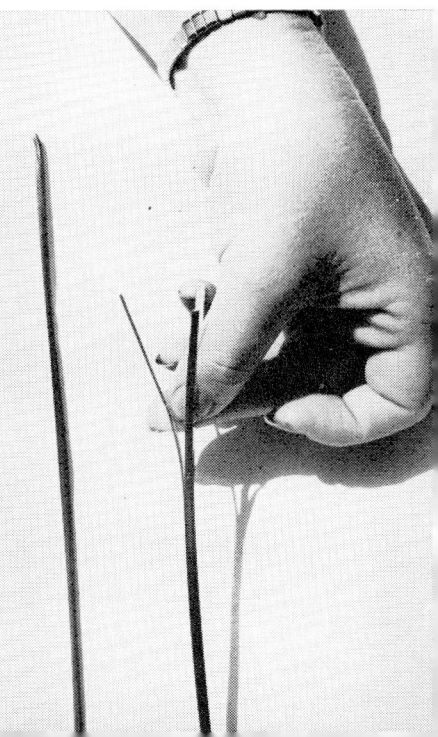

50. Hat made from *kuanaya* and pine needles

51. Splitting *kuanaya*

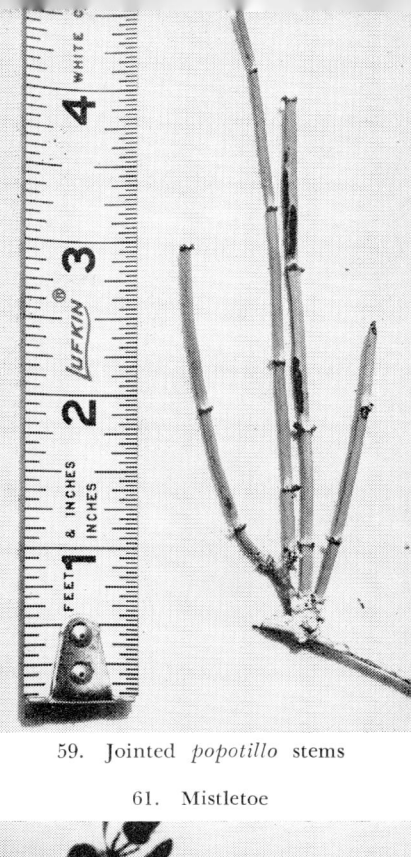

58. *Popotillo* plant

59. Jointed *popotillo* stems

60. *Flor de San Pedro*

61. Mistletoe

56. *Candelilla* growing in the desert

57. *Candelilla* stem with oozing sap

54. Tony Ashman examining squawbush shoots for basketmaking

55. Squawbush shoots being used in coil-binding basketwork

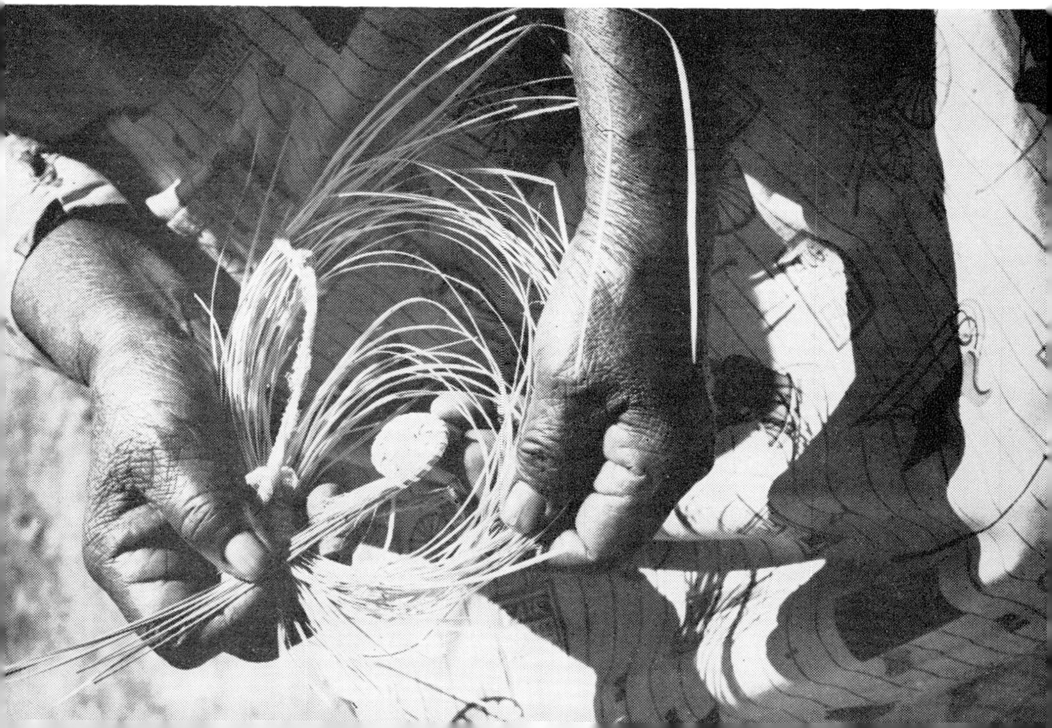

52. Juanita Nejo showing the stages of her craft

53. *Ocotillo* and adobe combined as building materials

62. *Ejotón* branch and bean pods    63. *Ejotón* growing

64. *Damiana* of Baja California   65. Elderberry blossom

66. *Té de la Sierra* leaves

67. *Té de la Sierra* plant

68. *Té del campo* plant

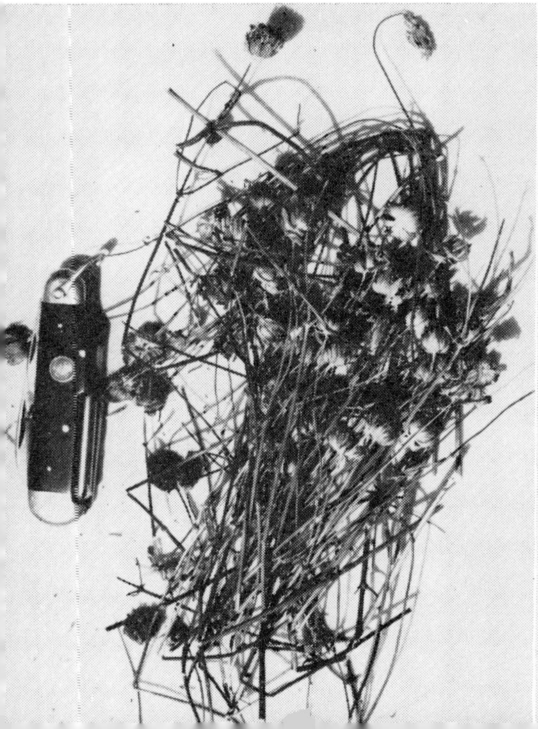

69. *Té del campo* blossoms

6

Beverage Teas

Several native plants are used in the brewing of various refreshing and pleasant drinks. These plants are found in the Southwest and in different areas of Mexico.

Damiana

Somehow, somewhere, the *damiana* (pronounced dah-me-ah'-nah) plant of Baja California became known as an aphrodisiac, and for many years tons of *damiana* were shipped from the seaports of southern Baja to destinations all over the world. Gathering the native shrub meant an increase in tortillas and frijoles to a good many families living in areas where *damiana* grew and jobs were scarce. But, after a succession of dry years and continued harvesting, the plants could no longer be found in abundance. A single order placed at La Paz from a large pharmaceutical firm for fifty tons of *damiana* was never filled. During rainy years, *damiana* is still collected commercially in southern Baja California and exported chiefly to New York. Its current retail price in La Paz is five *pesos* per kilo, or roughly twenty cents a pound.

A liqueur called *Créme de Damiana* is manufactured in Guadalajara and is flavored by *damiana* gathered from the interior of Mexico. In Baja, where the *damiana* is considered superior to that found on Mexico's mainland, the natives continue to gather the shrubs in limited quantities for home use as an aromatic tea. Its growth is pretty much confined to the territory of southern Baja where the shrubs first become noticeable a short distance south of El Arco, then continue to grow intermittently on down the peninsula to Cape San Lucas.

Tips of the branches, tiny leaves, and flowers, if the plant is in blossom, are all boiled together to make a delicious tea. I have drunk *damiana* on many occasions, and I heartily recommend it as a refreshing, flavorful drink. The accepted measurement for making most herb teas, *yerba del manzo* and eucalyptus leaves excepted, is *un porción de la mano*, or a small handful of twigs and leaves to a quart of water. *(See photograph no. 64.)*

Sycamore Bark, or *Cáscara de Aliso*, Tea

A pleasant beverage, this tea is similar to sassafras in both color and taste. The bark is chipped from the trunk near ground level, or from the roots slightly below the surface. The bark requires several minutes of brisk boiling to make a flavorful, red tea. *Cáscara de aliso* (pronounced cahs'-carah day ah-lee-so) tea is widely used in Mexico as a coffee substitute. Sections of orange or grapefruit peel are frequently cooked with the tea to add to its flavor. Young Indian women of the southern California tribes used to drink *té de aliso* as an aid to childbirth. *(See photograph no. 2.)*

Té de la Sierra

Té de la sierra (pronounced tay day lah se-err'-rah) is a beautiful green plant found growing throughout southern California and northern Mexico at higher elevations in the brushy mountains. The plant seeks shade and is easily recognized by the symmetry of its tiny leaves. The tea is excellent, but is regarded by some as having a slight medicinal taste. It makes a fine camp beverage for hunters and outdoorsmen. Branches and leaves from the root system upward are used; the tea is boiled, then allowed to steep. *(See photographs no. 66, 67.)*

Té del Campo

In blossom, *té del campo (pronounced tay del cahm'-po)* is a pretty plant. Its fine, long stems produce a prodigious quantity of bright blue flowers which, for size and shape, look a lot like clover blossoms. The flowers have an exquisite fragrance. This plant grows throughout the Southwest and is easily found in the spring and early summer. To me it has the nicest flavor and aroma of all the native teas. Stems and blossoms are used in the usual amount of a small handful to a quart of water. The tea should boil briefly. *(See photographs no. 68, 69.)*

7

Medicines of Rural Mexico

Trinidad Prieto, of Iagualica, Jalisco, Mexico, suffered severe stomach pains for twelve years, and as time passed, his condition grew steadily worse. As his pains became more intense, the more he avoided eating until finally he was subsisting entirely on a bland liquid diet. But even the drinking of liquids had its painful effect, so, with rigid self-discipline, he allowed himself to drink only the barest minimum of broth and milk in order to keep alive. There was no doctor in the area where he lived. "Trini" had no money and could not have afforded medical attention even if a doctor's service had been available to him. He was at that time in his forties.

Unfortunately for Trini, his many growing children all had healthy stomachs that constantly craved food in quantity, and in his house it seemed there was never enough frijoles and tortillas to go around. So one day sick in both body and spirit, he took his battered rifle from the wall and slowly meandered toward the mountains in search of a deer.

Although the Mexican sun blazed furiously, Trini had purposely taken no canteen. He knew only too well that if he took water with him sooner or later he would drink, and if he drank, he would be doubled over with pain and unable to hunt. After a long uphill walk, he saw a deer, chased it through the hot, brushy foothills, and killed it. While dressing out the buck, his overwhelming thirst compelled him to take a large leaf from a nearby plant and drink some of the fresh blood remaining in the lung cavity of the carcass.

Trini subconsciously braced himself for the recurrent surge of pain he knew would soon burst like an incendiary bomb in his stomach. But no pain came. He drank more blood, enough that he experienced the first comfortable feeling his stomach had known for months, but still there was no pain. His thirst was quenched and an almost forgotten sensation of well-being quietly settled over him.

Trini went on a blood diet. He hunted deer almost daily and each time he killed one, he carefully drained its blood into a bottle-shaped gourd. Whenever a goat or a cow was slaughtered in his neighborhood, Trini was on the scene with containers in which to save the animal's blood. What blood he couldn't drink fresh, he dried in pans, stored the powder in fruit jars and drank it later by mixing it with warm water.

In three months his ailing stomach no longer troubled him at all. Attributing his improvement to the blood diet, he lived to be a healthy eighty-odd years old before he died from other causes.

Trini's case is not an isolated one. The practice of drinking blood as a medicine for stomach and heart ailments is widespread throughout Mexico and has apparently been an accepted custom for a long time.

Juanita Nejo, who lives near the Pechanga Reservation

out of Temecula, California, and who is well along in her eighties, customarily takes dried deer blood, a heaping teaspoonful to a glass of warm water, whenever her heart is misbehaving. She tells me that she not only gets immediate relief from pain but also receives a sedative side effect that results from drinking deer blood which halts palpitation.

The deer's nose is also used by Indians and Mexican people as a remedy. The cartilage of the inner nostrils, as well as the black outer part of the nose, is assertedly the best vomit-producing medicine ever discovered. In cases of poisoning and other emergencies, these portions of the deer's nose are boiled in water and the solution is drunk by the victim while it is still warm.

I have a friend in Alpine, Texas, W. D. Smithers, who has probably compiled more information on the use of primitive medicines in the Southwest than any other person. He was born in San Luis Potosi, Mexico, while his father was bookkeeper for the American Mining and Smelting Company. Smithers went to Mexican schools until he was ten years old and learned to speak Spanish fluently. He then came to Texas with his parents and has since spent nearly all his life in the Big Bend Country of the Rio Grande. He became a photographer and newspaper correspondent at an early age and over the past fifty years has compiled the largest collection in existence of photographs (over fifty thousand negatives) and articles of that historical, scenic area.

He has always been particularly friendly with Spanish-speaking people, and because of his sincerity, he has been given information that few other individuals could have come by.

In one of his books, entitled *Pancho Villa's Last Hangout — On Both Sides of the Rio Grande in the Big Bend Country,* Mr. Smithers recently wrote:

To SAM HICKS: A compañero who also has an interest in Mexican life as this journal attempts to describe it.
W. D. SMITHERS.

With the express permission of Mr. Smithers, for which I am deeply appreciative, the following paragraphs are taken from a chapter entitled, "Nature's Pharmacy and the *Curanderos*" in the above-mentioned book. Mr. Smithers' experience and background have been unique, and his statements are based upon that experience and on accounts told him. Fascinating though they are, none of these experiences with folk medicine have been verified scientifically.

> In the making of some of today's scientific medical drugs and remedies, some of the native plants and products of the Southwest are being used. The curing and healing powers from them were known to many of the older Mexicans and Indians as their grandparents had used them hundreds of years ago. As late as the twenties, many of the Mexican men and women over forty years of age that lived in the lower parts of the Big Bend District knew of the curing powers of the various plants that grew wild in that region. Some also had in their gardens a few herbs that originally came from Mexico.
>
> Those that knew of all these plants and had attained a reputation for the curing of the sick were known as *Curanderos* (the men) and *Curanderas* (the women). That means a healer but not a doctor. Those that had become well known for their abilities were called *Brujerias,* which has a meaning of Wizard, or as we refer to a specialist among our doctors.
>
> It was not a profession that these people followed, as they made their living as goat herders, gathering and selling chino grass, working for American ranchers, or other kinds of work. In most cases the women were superior, as they acted not only as a doctor, but also as a nurse. They charged no fees for what they did for the patient, and besides the treatment, they often fed them for several days

while they were in their care. This is a true demonstration of the generosity of Mexicans to share what they had, even if they knew they would have to do without certain foods after their guests and patients left.

What these people knew of their skills had been passed on to them from many generations. Few were able to read or write. It is not known when our food, drug, and medical laws were enacted, but if they were on the books in the early twenties there were many violations of them. They not only made the medicines, but they diagnosed the patient for his ills and treated them. They were both doctors and pharmacists without a license, but from results of numerous cases they treated they were capable and did much for many people.

The purpose of this story is to tell that Nature had the plants available that could cure, and there were a few who knew all of them and for what each would be used. Also that today's drug manufacturers are using some of these plants to make medicines. There probably would be more of them used if they could reach the laboratories in the same condition as when they are gathered. Maybe some day there will be laboratories near where the plants grow, as it is believed that in the future many of the plants will be of value to mankind.

Also, it is the desire to record as nearly accurately as possible the prescriptions that the Mexicans made up to treat the patients. Some of these have proven very successful, from my personal knowledge, but of some there are doubts, as this writer is not a medical authority and did not know if the patient had the ailment for which he was being treated, but the odds are in favor of the *Curandero*.

These people were very fond of all plants; each home had two gardens. One was of various vegetables, corn, beans, and melons. This was near a dry creek bed, a short distance from the house. With a dyke and ditches that they built the flood waters of the creek were diverted into their garden after the rains that made the creek flow. The other garden was at their house. This was for flowers, herbs and some of the native plants that they used for medicines.

These had been transplanted, but others would not grow in the garden so they had to be gathered where they grew. The flower garden had to be watered by the children carrying buckets of water from the river. What they grew in those two gardens was not only some of the ingredients for the prescriptions, but a number of the vegetables were important for the diet that the patient was often kept on for several days.

It is believed that much of the success in their *remedios* (remedies) was in the knowledge of the proper food to help the medicine correct the ailment. The patients were cooperative as they had confidence in the *curanderas* and they followed the instructions given. From personal treatments and from knowledge of others, it appears that the women were better than the men, but both appeared to be in charge of the case, working in harmony, both agreeing on how to treat the patient.

This belief that the women were better might be biased for, during the years that this writer lived among the Mexicans, as a critically ill patient three times, two of those cases, and they were the most serious ones, it was two different women that I am grateful to. One was a bed case of *Tefo* (typhoid fever), and the other was a severe sunstroke. My third hospitalization was for yellow jaundice, in the care of a goat herder and his wife. Their medical knowledge and proper food I needed had me cured in a week.

When a patient went to a *Curandero* he had to go to him early in the morning, before sun-up, or have his illness diagnosed by the woman. Most patients arranged their visits late in the afternoon, planning a stay of several days. When the patient's ailment had been diagnosed and the decision made as to what would cure him, then the prescription was not filled, but it had to be made up after the necessary plants were gathered.

Some of those plants grew nearby, but some were several miles away. Usually one of the boys was sent to gather the needed plants which were always prepared and made into the medicine by the woman with her household utensils. When the plants or flowers were put into the *cazuela* (clay pot), to be boiled, they were put in just as they were gathered. Nothing was done to destroy Nature's touch. Most all of the plants that grew in that area were used for medicines, foods, or uses around their homes. Some were used for more than one ailment, and they knew the proper amounts to be used to compound for the prescriptions and how much the patient should be given, also how often.

A few of their remedies this writer cannot vouch for, except that they were told by persons who knew that they were successful, others were seen when used, and some were used when I was a patient. One was for a severe case of sunstroke, *pico el sol*, as they called it. I was alone when I passed out, but recovered and was able to reach Johnson ranch.

This was June, 1930. Mr. Johnson's car had broken down and he borrowed my Dodge to go to Terlingua, 32 miles away. About 1 p.m. Mrs. Johnson and I had just eaten our lunch when a Mexican boy rode up to the house on a burro to tell us that Mrs. Harloe had been shot. The boy was so excited he could give us no facts except that she was shot. We had an air force medical kit in the house as Johnson's ranch was an army landing field.

Not knowing when Mr. Johnson would return with my car, I took the kit and started to Mrs. Harloe's house, which was about 3 miles by the road, but I took a straighter course, saving about 1 mile. The mistake I made was setting my pace at a trot in that heat that was at least 110. Just before reaching her home I felt a slight dizziness, but continued on, reached her house hot and exhausted, and soon saw that my hot, fast trip had not been necessary.

Near her home was the Holquin settlement. Mrs. Holquin had gone to her aid. It was only a flesh wound about 3 inches long on the lower part of her right arm. Her children had been playing with a 22 caliber rifle, and she told them to put it down. They obeyed but put it on the bed. Mrs. Harloe picked it up by the end of the barrel, the hammer caught in the quilt and discharged.

Old Juana had stopped the bleeding by applying soot to the wound, then spider webs. That was their standard remedy to stop bleeding, even serious knife wounds of those wild dance fights. Mexicans never destroyed a cobweb unless they needed it for a wound. They knew where every one was in the house or outside.

Juana had just completed washing all of Mrs. Harloe's arm when I arrived, and she turned her patient over to me. The medical kit was opened to give her a couple of Aspirins, but there were none. Someone had taken the bottle out and failed to put it back. There were several other bottles of pills but I did not know what they were for so all that I could do was to swab some iodine on the wound and bandage it. I decided to go back to get the aspirins, but to go a route closer to the road so that I could see Mr. Johnson in the car if he passed. I did not trot, but as I am a fast walker, I over-exerted my exhausted condition.

The Johnsons' ranch was on a large mesa and as I was climbing it I became weak and dizzy, but kept going, reached the top, then suddenly I blacked out and fell. How long I was out I never knew, but from the greasewood that I knocked down when I was rolling around, it must have been ten minutes. My arms, face and clothes that were wet with sweat were covered with mud. I had numerous cuts and bruises.

When I came to I was in a sitting position. I could not recall getting into that position, but shortly I rose to my feet. At first my mind was not clear enough to get my bearings, everything looked out of focus, but shortly the distant south end of the Chisos mountains became a clear image, then I knew just where I was. I was about 500 yards from the Johnsons' house, and about 100 yards from the road. I realized that I did not have my hat on, so I started looking for it, then I heard the motor of my Dodge start up. The hat was not found as I headed towards the road and the Johnsons saw me.

Their first intention was to take me back to the house, but Mr. Johnson thought that old Juana would be the best doctor. In the shade, at the east side of her house, she placed a couple of goat skins on the ground for me to lie on. She sent the children to gather some of the plants known as Wild Tobacco (Nicotina Clauca), *Hojo de Igera* they called it, also *Piel de Elephante,* which means elephant skin as the roughness of the leaves resembles elephant hide. This plant grew abundantly along the Rio Grande, but it was said that it was not native to this area, originally coming from South America by way of Mexico. The flowers were trumpet shaped, yellow, and they were the favorite flowers for the humming birds.

While the children were gathering the plants, Juana took from a jar some dried sunflower seeds. These were from the large sunflowers that they cultivated in their garden. They called them *Mira Sol,* which means looking at the sun, as this flower does just that all day. The seeds were placed in her *molcajete,* which is a stone bowl they use to mash up chile by pounding it with another long shaped stone. The mashed-up seeds were then mixed with cotton seed oil *(aceite de comeder)*, oil you can eat.

That mixture was spread on the wild "tobacco" leaves, some of them were 6 inches long and 4 inches wide at the widest part. They are a heavy, pliable leaf. Juana covered all my forehead and temples with these leaves and told me that another treatment would be given me in the morning, as the brew had to be boiled, then let to cool over-night and would be poured over my head next morning. She let the first batch of leaves stay on about one half hour, but during that time she doctored up the cuts and bruises I had on my face and arms. This she did with a salve she already had mixed up. She told me it was made by rendering the fat from near a goat's kidneys with pulverized dry blades of the Palmas (a species of the Yucca-Elata).

Juana kept asking me if I had any pains in my heart, but I did not. I felt weak and sore, a little sick in my stomach, but most of my discomfort was in my head. Within a few minutes after she had applied the leaves I felt some relief. She let the first ones stay on about half an hour, and as she took them off she said that they were *muy colorado* (very red) but I did not see them. As fast as she took one off, she put a fresh one on.

It was about sun down when she took the leaves off and told me not to wash that night, but to come back early the following morning for the next treatment. I felt much better, but I was sure weak and sore, felt like I had been given a terrific beating. Mrs. Johnson, Ada, prepared me a rich, good beef broth from cans of her beef stock that she canned when they killed a beef. The next morning, after breakfast of oatmeal and soft boiled eggs, I went back to Juana's house for the second treatment.

They had probably been boiling flowers all night from the amount of the brew that they had ready. This was not to be taken, but it was prepared to treat my head with this solution, a sort of shampoo. It was made by taking the entire flower and a part of the stem of the small native sun flower, *Flor del Sol (Helianthus Annuus)*, boiling them for at least one half hour. The brew was then poured off to cool over-night.

In a sitting position with my head down, she slowly poured about a quart of the brew on my head from my neck to the brow. She allowed me to sit erect about half an hour, then repeated the treatment. After the second treatment of pouring the sunflower brew over my head, she again allowed me to sit up and let the juice soak into my head, then, still sitting erect, she poured on small amounts, some of the liquid, and with her fingers gently massaged all of my head for nearly half an hour. That completed her cure for me.

At the time of this sun-stroke, I was a veteran of 15 years, most of it in this region. Five years were with the Cavalry and pack trains. On numerous occasions I had seen the medicinal men care for a trooper that was knocked out by the heat. They were fortunate the medics were there to care for them. It is a terrifying sensation to come out of a coma when you are alone out in the greasewoods, many of them crushed down where you have rolled around on them.

Exactly how I fell was not determined, as there were

cuts and bruises on my face and a bump at the back of my head. My hat (later found) was ten feet from where I fell and rolled. This happened 30 years ago, no ill effects of it have been felt, but it has been wondered many times how long I was passed out? There were two mistakes made by me that day, and I knew better; one was that I should not have set such a fast pace in that heat. I am sure that, had the Harloe's home been a little farther I would not have reached it, as I was nearly exhausted. I was craving a drink of water, but I had sense not to take one, but before I started back for the aspirins I did, and I am sure that it was the water which was partly responsible.

. . . .

The best life saving plants are the cacti. Not only are the fruits of them eatable, palatable, and contain much nourishment, but the plant itself is also. The round species that grow on the ground are the best. The top is cut off by inserting the knife blade into the side, a couple of inches from the top, then making a horizontal cut around it. The top is lifted off, chunks of the center are cut out and chewed. It is surprising how refreshing and pleasant the juice is. What is in the juice not only ends your craving for water, it also replaces some food value to your body. The pulp of the cacti is only chewed, but the fruit that many of the cacti have are eaten, and from them you get the best food value. The young, tender blades of the prickly pear cactus are eatable, raw or toasted over a fire. The cabbage-like head of the sotol, when put into a hot fire and toasted, makes a good meal.

Much can be learned from watching and studying the animals and birds in the arid regions of the border country. Notice the species of cacti that animals have gnawed into to get needed moisture. Some may have been rodents, skunks, foxes, coyotes, and deer. The specie that they select is the best one for you. The best way to locate where water can be found is to watch the doves. When you see numbers of them flying in the same direction towards a canyon or a draw in the mountains, follow their course, and you will find a spring or a *tinaja*. Animals' trails will also lead you to water. Wild life make good teachers, you can learn much from them and in a rough country to make it easier you cannot learn too much.

During the border trouble days of 1916 to '20, many of the soldiers that were stationed along the border were from states more distant. Most of them could not understand how anyone could make a living in such a place. They were hard to convince that nearly everything that grew could be used for food, medicine, animal fodder, or for household use. It is true that this country had no resemblance to the green meadows of their home states, but on top of the rough mountains there is an abundance of the best grass in the world, called Chino. For horse feed that grass is equal to both hay and oats. Many of the weeds and bushes that looked worthless were good feed for goats and deer.

It was the officers and enlisted men of the Medical Corps who took an interest in and studied the natives, especially the *Curanderos*. The semi-primitive methods of these people used naturally were of interest to the army doctors, they knowing the history of medicine.

A remedy that they claimed was very good for dysentery was made from carbon (charcoal), but it had to be made from a log of a mesquite tree. Usually the *Curandero* (healer) had a piece of it ready made to fill the prescription for the patient, but the remedy was made this way:

A hole was dug in which a fire was started, then a green mesquite log was put into it, then smothered and allowed to burn into charcoal. It was then pulverized into fine powder, mixed in water. A glass of this mixture each day for three days had you cured, according to their testimonies.

This treatment was never used by this writer, and when told to some medical men, they expressed doubts that it could cure dysentery, but did they know what was in the mesquite wood?

When they treated a person for a heart ailment it was

70. Rural Mexican garden for flower and herb growing

71. Spider web carefully preserved for use to stop bleeding

72. Dried buck deer blood taken for stomach and heart ailments

73. Deer-nose, dried and powdered, for emetic properties

74. Scaly red eucalyptus
75. Smooth white eucalyptus
76. Shooting *guatamote* shaft
77. *Guatamote* for medicinal tea

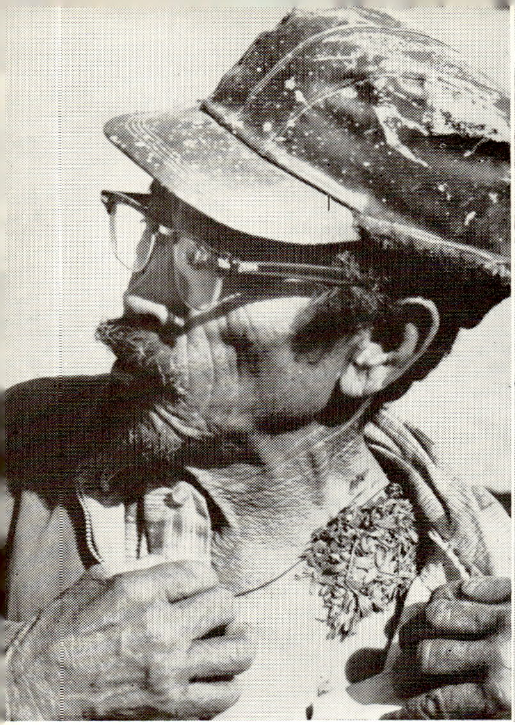

78. Gorgonio Fernandez showing *gobernadora* poultice    79. *Gobernadora* used for tea

80. *Higuera cimarrona* figs    81. *Higuera cimarrona* tree

necessary for some one to go and kill a deer. While that was being done, one of the several shedded deer antlers that they always had around the house would be ground up into powder. The old dried bleached horns were much better. They were called *cuerno de venado*. They pulverized them by rubbing one on a horseshoe file or a nutmeg grater. They needed about two teaspoons of the powder.

While the deer was still warm, they drew about a half cup of the blood, added some warm water, and the horn powder, stirred it for several minutes. The patient was then given this concoction to drink in small sips slowly. Later, the patient was given another cup of blood without the horn powder, and less water in it, then, still later, some of the blood undiluted.

As this treatment was never taken by me, this is from statements of several that had great confidence in this cure. The blood of a buck was the best, but a doe, and even a goat was also used.

When one has a bad case of yellow jaundice and is more than a hundred miles from a doctor or a drug store, also if he is alone, camped out in the Big Bend District, he needs someone to help him. That was another time I was cured by the *Curandero* and his wife. That was in 1924, about 14 miles below Castolon and 118 miles from Alpine, where the nearest drug store was. There was at Terlingua, 30 miles away, the doctor who cared for all the personnel of the mines, but only one mile from my camp lived Alehandro Domingos and his family who was a *Curandero*.

All my camp was loaded into my Dodge roadster by daylight. When I arrived at their home, as the sun was rising, Alehandro was just about to start off with his goats. When I shook hands with him and said that I was *"poco malo,"* a little sick, he looked at me and said, *"Si, señor,"* yes sir. I was as yellow as a lemon, anyone could have seen that I had yellow jaundice and I felt that I had several ailments. I had been trying to fight it off for three days and I knew that I was getting worse rather than better.

Alehandro sent his oldest boy with the goats and we

went into his house, told his wife that I was very sick and would stay with them for a week. He then said he would go and get the plants to cure me. He returned with two different kinds, only one of the Ojazan, Tar or Black Bush (*Flourensia Cernua*), but several of the Popotillo, Kidney weed (Ephedra Trifurca).

Mrs. Domingos had the water boiling in a clay *cazuela* into which she put the Tar bush, leaves, stem, and the stalk, broken up into small pieces. It boiled for about half an hour, then was set off the stove to cool. Later the brew was poured into a glass jar, about a pint of it. This was a laxative, and they had the knowledge of how much of the brew to give the patient. I wondered if I had to drink all of it, but later only about a third of a glass was given me. It was about 10:30 sun time (we went by the sun, not a watch) when I was given the first prescription, it was luke warm, but in my condition I could not detect any taste to it. This was the first time I had ever taken this medicine, but at later times, when again I took it for a laxative, I recall that it had a tangy acid flavor, a little bitter, somewhat like the taste of SSS tonic, but a stronger flavor. It affected me within three hours.

While the dinner was being cooked, over one of the four openings of the wood stove, some of the Popotillo was being brewed. This was given me later to correct the disorders of the kidneys and liver. Mrs. Domingos, Alicia, was preparing two meals, one for her family and a special one for me. The diet that she kept me on for a week was the main part of the cure. During that week, I ate nothing that had any grease in it.

The noon meal consisted of a salad she made from the partly grown joints of the prickly pear cactus. These were cut up into cubes for the salad, the tender part of the thorny cactus. They were about ¼ inch thick, 3 inches wide and 5 inches long. They are very good for salads or dipped in a batter and fried like egg plant, but fried food was off my diet for more than a week. With the cut-up cactus was added some mint leaves from her herb garden, a few pieces of onion stalks and some of the juice from the fruit of this cactus, and a little of the fruit cut up and all mixed. It was a good salad. The main meal was a meatless stew made of squash, okra, small onions, green beans, turnip greens, and the small ends of very young ears of corn cut in slices. She told me to eat the cob part also.

About mid-afternoon she gave me half of a water melon. Most of the melons that they raised in that area are very small, appear to have been pulled when only part grown, but they are ripe and delicious. They are pulled before sun-up, placed in the shade of their home, and covered with a wet sack. When eaten in the afternoon, they are not cold, but are cool, as Nature meant them to be eaten.

Before supper, about a quarter of a glass of the brew she made from the Popotillo plant was given me. This was taken twice the next day, then once each day the following two days, but a milder brew. On the morning of the third day, I felt much better. At sun-up the fourth day I went with Alehandro and his goats, made pictures and learned much of the life of a goat herder, but he made me go back to the house at noon.

The food during the week I was in their care was about like the first meal except Mrs. Domingos added to the stew pieces of catfish that they caught in the Rio Grande, and on the fifth day the stew was made with the breast of a chicken cut up into small pieces. Also served were tender young mesquite beans boiled with a few pieces of cinnamon bark. Various greens were a part of the diet, turnip, spinach and mustard greens boiled with the green stalks of onions. Also served were large ears of corn partly boiled, then roasted over an open fire. Half a water melon was a treat each mid-afternoon.

If on the seventh day there were any symptoms of Yellow Jaundice in me, they were not felt, nor could they be seen. Most of each day had been spent with the children gathering the vegetables, carrying water from the Rio Grande, gathering firewood and going to nearby interesting places to see and make pictures of. Those kids, too, helped me to make a speedy recovery. They probably were sorry when I was cured and left, as I have always had a habit of chewing tobacco and with it chew a stick of gum, I still do. Many ball players do. The kids had lots of gum that week I was a patient of their mother and father.

When the time came for me to depart, diplomacy had to be used to reimburse them for what they had done for me. All that they expected was that I show by my actions that what they did for me was appreciated, but I wanted to do more. In my grub box was a two-week supply of food, mostly canned goods, coffee, bacon, and flour. Telling them that I planned to go to Alpine, then to San Antonio (but not when), there was no need for me to have those items. This was the only time that I ever drove off in the Big Bend country with an empty grub box, but it was again full when Castolon was reached, including a new can opener, as they had never owned one.

Some of the many other cures and treatments that the *Curanderos* practiced have been told in other chapters. A few of their treatments are identical to those that were used in some American homes in the pre-twenties. A cold cure was: place a large slice of an onion on a saucer, cover it with sugar, then let it stay in the cool air over night. In the morning, eat the syrup and part of the onion. Mexicans also used the sweat remedy, a hot lemonade or a hot tequila toddy, then under the covers, sweating out their cold.

To relieve swelling of sprained ankles or wrists they made a poultice of sun flower seeds, as Americans do with flax meal. Yellow laundry soap mixed with sugar made a poultice to draw the pus from sores and boils. Poultices were made from cacti, some from seeds and plant leaves mixed in oil made from goat or deer fat. Various ones were used for burns, bruises, sprains, or other ailments. It seems like they made one for any kind of a pain, and what they made up appears to have been the cure.

A poison treatment was to make the patient vomit as

quickly as possible by drinking a large glass of warm water in which some lard had been dissolved, then some vinegar added. This was followed by a glass of sour goat's milk. The patient was kept on a diet for several days of vegetable soup, some greaseless cooked vegetables, boiled mesquite beans, and water melon.

There was one case of food poisoning that an entire settlement of fifteen persons, men, women and children took this treatment. Two of the women were *Curanderas,* but their condition was so serious that a doctor from Alpine, 123 miles way, had to be sent for. This happened in 1933, at the Holguin home, and the two homes in Mexico across the Rio Grande. Every member of the three families were stricken except the oldest Holguin boy, Brijador, who was away that day. When he returned that night he found all his family and those across the river very sick (*muy malo*) and scared, as they did not know what had caused it. Brijador rushed to the Johnsons for help.

What had happened was that one of the Holguin boys had been taking lime from the Johnsons' tool shed. This was kept to mark the landing field. Lime was used by the Mexicans to soak the corn in before they made it up into tortillas. The boy made a mistake and took from a similar looking sack some poison that was used to dust cotton plants. The corn was soaked in the poison, tortillas made, eaten, and nearly killed all fifteen.

This was a very unusual medical case in this area, and took a little detective work to learn what was the cause. The doctor had to make almost an all night drive over those rough roads, but they were all alive because they had taken the lard and vinegar treatment. They were one bunch of people who were glad to see a doctor. All recovered.

. . . .

During the Mexican border trouble days many of the bandits were wounded by the U.S. soldiers, but they got away, sometimes after they had been shot off their horses.

The *Curanderos* in Mexico must have had a busy time curing the many wounds that were inflicted by the troopers. Several of the bandits whose identity was known, were wounded several times, but managed to get back across the Rio Grande. There was one that every cavalry man, Texas Ranger, and Custom Mounted Officer made every effort to shoot dead off his horse. They did wound him three times, but that bandit lived to an old age and died a natural death in Mexico.

During the Escobar revolution of 1929 in Mexico one of the last battles was at Jimenez. The Mexican Federal troops turned the battle into a rout. Some of the revolutionists did not stop retreating until they reached the Rio Grande, more than 200 miles away. Some had been wounded, a few were in bad conditions from lack of medical care. As they were still in Mexican federal territory they were concealed in the mesquite brush near the Rio Grande. The *Curanderos* and their aides kept them in food and treated their wounds daily for more than three months. One had three bullet wounds in his abdomen and gangrene, but he lived to be a resident of that area later. He refused to talk about it and the *Curandero* who was believed to have cared for him would not admit that he had cured the bandit.

Mexicans were strange people when asked to give information that might involve someone and get them into trouble; even to a trusted friend. The *Curanderos* were like doctors, and would not discuss their patients' cases, but the desire was to learn how they treated those serious bullet wounds, some inflicted by the American soldiers and others by the Mexican federal troops. The best guess is that they used poultices made from cacti, plants, and leaves, some that had been brewed, taken internally to build up their bodies to help cure the wounds. The food must have been an important part of the cure. What percentage of the casualties survived never was learned.

Damiana (Damianita), *Chrysactinea Mexicana*, was one plant that was known as the plant for the *mujeres* (women). Few of the men would admit for what and how this plant was used, and the women, in reply to questions about it, would say *no sabe* (don't know). They all knew but did not want to talk about it.

This plant is a low-growing bush, usually up to one foot high. It has numerous dark green cylindrical leaves which make a strong scent when crushed. The flower stalks are about three inches long, emerging from the cluster of stem leaves as small as daisy-like yellow *(sic:flower)* with yellow center. Each flower cluster is about one inch across.

Reliable facts on how this plant was used are few, but it was learned that, when properly prepared and used correctly by one with the knowledge of it, that it was a cure for some female disorders. It is said that it is a dangerous remedy to use unless by one that is an authority on it. Other reports about it are that it was used for abortions, that being illegal no facts could be secured.

It is believed that, in some of the larger towns on and near the border, there are some quack *curanderas* that secretly use this plant on pregnant women, also treat other patients with various plants for other ailments. All of their treatments are illegal; practicing medicine without a license, and a few of the lesser offenses are found out by the authorities, such as when a person pays them for a cure for an ailment and is not cured. The serious cases, when the Damiana is used, never are known, for all involved will not talk. The saying was that, "It cures or kills."

Without positive evidence, one case where Damiana plant is believed to have killed a young woman will be told. Juanita (not her real name) was 18 or 19 years old, and appeared to be in perfect health. She was involved with a married American ranchman. Later, the rumors were that she was pregnant. She died suddenly. One rumor was that she had taken, or was given a poison, but it is believed that she gave herself a treatment of Damiana. Her mother was one of the best *Curanderas* in that area, therefore Juanita had a knowledge of the native plants. It is certain that her mother did not prepare the treatment, or knew her daughter took it.

How Juanita took this treatment was not known, but it was believed that she drank the brew she made from the plant. That probably started the report that she was poisoned. The only reliable information obtained on such a treatment was told by the woman who knew a pregnant girl who took it in a town in Mexico. A blood transfusion set was used to inject the brew into the womb, according to the girl's story told to the woman that told this writer.

How many such treatments have been given in the Southwest will never be known. Authorities are powerless to obtain evidence for conviction. The quack *Curandera* lived in a house with her family among many other similar houses. She had occasionally young women visitors for several days (supposedly relatives from small towns about 200 miles away), but that is the way of most Mexican families. They have frequent visits from relatives. The most accurate estimate of the fee was $50.00.

. . . .

Popotillo, *Ephedra Trifurca,* is known by several English names and the medicine made from it is used for several cures. Common names for it are kidney weed, clap (gonorrhea) and straw weed, for it looks like a bunch of straws, smaller than the candelaria plant, but similar to it. The stems or stalks with very few small leaves they have are boiled into a brew. The *Curanderos* knew how much of the plants to boil to make the required strength for each disease.

When properly used, this was very good for kidney and liver disorders. It was this prescription that cured the yellow jaundice this writer had, and in only one week. When used to treat a patient that had a venereal disease, the plant was brewed in the correct strength to be taken internally, also other stalks were pounded up to make a poultice. As this treatment was never used by this writer it is told from statements of those who used it, or saw it used, and they claimed that it was a good cure.

Snake bites (rattler), needed quick aid, generally given

by the victim if he was alone. First a torque was applied above the wound. With a knife a big X-shaped gash was made at the wound to let the blood get the poisonous venom out, then, if the wound could be reached with the mouth, the rest of the venom was sucked out. With someone to aid, this was a simple task, but if alone, you had to do the best you could. To get the venom out as quickly as possible was what was needed to be done.

Few people know it, but often the shock and terror of a snake bite is nearly as bad as the bite. When a snake has struck, it is then too late to become frightened, it weakens the body and the venom will have more effect. A calm mind makes the body stronger, which is needed if the snake succeeded in getting the venom in. Sometimes, when struck by a snake, its fangs do not penetrate deep enough to inject the venom. Those are some of the cases where the shock was worse than the bite.

The rest of the Mexican snake bite cures were to give the victim a cup of half warm water and goat blood to drink slowly. Later a cup of the pure blood. This they believed replaced the loss of blood when the gash was made. On the wound a poultice was placed, some used one made from the mashed-up stems of the Popotillo, others made from the Prickly Pear leaves or of the dried Yucca blades mixed in goat tallow.

In south Texas, where there are many rattle snakes, the catching alive of the snakes was the occupation of many Mexicans. They were sold to dealers in various towns for 35 cents per pound. The dealers bought all they could get as there was a big demand for them. They were shipped all over the United States to medical laboratories, zoos, colleges, sideshows of circuses, and to several cities where there was a large Chinese population, as they are very fond of snake meat.

Most of the professional snake catchers and dealers were immune from a snake bite. They had been bitten so many times in the years of their handling them. Of several that were seen to be bitten only one, an old Mexican about 70 years of age, was bitten on his right index finger, about where he had been bitten three previous times, but this time, he told me the next morning, it made him sick in the stomach. He was again catching more snakes the next day.

. . . .

About the only use the Mexicans had for rattle snakes was by rendering the oil and grease; this was rubbed into their joints when they had rheumatism or sore muscles; but the severe cases which are now known as arthritis, the patient was immersed in a tub of a brew made from the Gobornadora (Creosote) *Larrea Divarricata*. This plant was often erroneusly called greasewood. It grows abundantly throughout the entire Southwest, and is used for several other purposes besides its medical value. In 1857, when the army was using camels to transport supplies to the troops in this area, the camels preferred this plant for food; they relished it. It grows two to three feet tall and has many green heavy leaves.

The brew for the treatment was made by boiling large amounts of the leaves and parts of the upper stalks, then poured into the largest washtub, the patient put in to soak and be sponged over all the ailing parts for about one-half hour. The brew was as hot as he could stand and a treatment was repeated for three days. From reliable reports it was a good cure for what is known as arthritis. The roots of this plant were boiled in small amounts of water to make a brown-colored dye which was used to color parts of the wool used in weaving blankets.

For this same rheumatic ailment they also used a different plant, made into a brew and used about the same as the creosote, but most preferred the creosote. The other plant was the Lechuguilla (common name) *Agave lechuguilla*. It is a cluster of heavy blades about 12 inches long, similar to the Yucca, but has no stalk. At the end of each blade is a needlepoint thorn, and, unless when walking among them they are avoided, your legs will be penetrated by them. Besides the pain of the thorn in the flesh, it leaves the film from it in the wound and this causes an irritating and painful feeling for a couple of days.

Mr. J.E. Casner, a rancher and General Motors dealer in Alpine, and a man of recognized scientific ability, has worked with scientists during the past few years, and a definite quantity of cortisone has been found in the lechuguilla. The entire lechuguilla plant is boiled to make the formula. This plant has been shipped in carload lots to a laboratory in the Midwest, where a medicine is being made to treat arthritis. When proven successful, it will be made to be injected, to be used as a liniment, and in powder form to be taken. However, the burdensome cost of freighting the green plant has created a problem. Maybe the old *Curanderos* knew that there were good medical cures in that plant, but did not know how to extract its best medicinal qualities.

The lechuguilla, beside being a medicinal plant, was used for several useful purposes. The blades were stripped into fiber, in about ½-inch strips that made an ideal cord to tie the bundles of chino grass. When stripped into finer strands they wove it into ropes, mats, brushes, and various other items they needed in their daily life.

Also, Mr. Casner has experimented extensively with the utilization of the durable fiber of the lechuguilla plant for making scrub brushes and insulation fiber. The hand stripping of this fiber, as it is done on a commercial scale in Mexico, is prohibitive in the United States because of labor costs. Casner and a Mr. Hannold (now deceased) built a machine for stripping the fiber from the plant, but this machine has never been satisfactorily perfected. If and when such a labor-saving machine is perfected, millions of tons of the plant can be profitably harvested in the Big Bend area of Texas. And, this will represent a twofold gain: several million acres of ranch land will be cleared of a plant that has no food value to domestic livestock, and the fiber and medicinal properties of the plant can be marketed.

A plant Mexicans called San Sipriano (sage brush) was made into a brew, taken for a period of three days, often longer, to cure diabetes. This cure cannot be verified by this writer, as he never talked to anyone that had taken it; but from the statements of several who knew of others that did, it could have been a good cure. It is known that the Indians of the Southwest used this plant as a medicine, but for what is not known.

For a blood tonic a bird was used. It was the Road Runner, Chaparral bird, or the Paisana, which means a good neighbor. This bird travels on the ground most of the time, but can fly as well as he can run. It is said that he can outrun a horse. Its favorite food is snakes, lizards, mice, and it is very destructive to birds' nests, eating the eggs and taking the young birds to her young.

The bird was prepared by killing it without losing any of its blood, then it was taken to the river bank and completely covered with wet clay all over the feathers. In a hole dug in the center of a large fire the clay-covered bird was put in, covered with hot ashes and fire, to cook for about one hour. When removed from the fire, the baked clay was taken off in pieces and the feathers came off with the clay. The meat looked very appetizing, but the thought of it being cooked without being dressed spoiled the desire to try it. The entire bird was the prescription for the cure. This bird's body is about as big as a small pigeon.

The list of plants that were used by the *Curanderos* is a long one, and they knew exactly what was the best for each ailment and of what strength to make it up. In some instances, considering the conditions that the medicine was made up and the hospitalization of the patient, they were lucky that they survived. In the house where this writer was a patient with Tefo (typhoid fever) for about three weeks, a hen laid an egg on the foot of my bed in the morning of the day that my fever reached the crisis. She continued to lay her eggs each day while I was bedridden, this I saw; but the Curandera swore that the morning my fever broke was the first time. That made her happy, for she knew that I was going to recover. Mexicans are strong believers in omens. As I recovered, those fresh laid eggs were served to me and they were a treat as my diet had been burro milk, but they could not have given me a better food.

In this statement I may be wrong, but it is believed that the *Curanderas* or *Curanderos* were better qualified to cure adults than small children, but it is believed that they lost more infants and children up to about five years than they did the older ones.

In the Big Bend District and other parts of the Southwest, except in Mexico, the old *Curanderos* are now few in number. Their children have no desire to learn all those primitive remedies, for even those who live at the farthest parts can reach a doctor's office in less than three hours and they prefer to have the doctor give them shots, pills or better tasting medicine than what their grandparents used to make up. Mexicans are great believers in going to a doctor or a *Curandero* when they think that they should, often when they don't need to. All the relatives and close friends are very sympathetic of the one that is sick, often it seems that they suffer as much as the patient does.

It must be emphasized that the Spanish word *Curandero* means a healer, but not a doctor, and they did not try to take credit for being a doctor, but to one in need of medical care they were a good substitute for a doctor. In their regular work that they made their living from, they observed all the various plants that some day they might need for a cure. When one was needed, maybe a burro ride of several miles had to be made to get that plant to fill the prescription for the cure. They were willing to do that for anyone needing care.

The above paragraphs, taken from "Nature's Pharmacy and the *Curanderos*" by Mr. Smithers, are quoted exactly as they are written except for the fact that some intervening paragraphs are omitted. *(See photographs 70, 71, 72, 73.)*

8

Alamo, or Cottonwood, and Other Trees

The trees of the Southwest are very vital to its peoples. In an arid desert area, a tree can provide lifesaving shade. Even under more moist conditions, the pleasures given by a tree can be immeasurable. In addition, many of the trees of the Southwest have healing properties and are widely employed throughout much of the United States and Mexico for these properties, as well as for more common uses such as fuel and building materials. *(See photograph no. 84.)*

Eucalyptus

For over twenty years La Botica Central of Chihuahua, one of the largest pharmaceutical companies in Mexico, sold by mail order white eucalyptus leaves for thirty pesos per kilo. The price is higher now, but the consumers' demand remains the same. A kilo of white eucalyptus leaves contains an untold quantity of cures and is still the usual Mexican family supply for one year.

Since the introduction of these trees into the United States and Mexico, a wide variety of ailments has been successfully treated, or cured, through the continued use of their medicinal properties. Eucalyptus teas and oils are still being used in the Southwest for the regular treatment of malaria, bronchitis, tuberculosis, asthma, kidney stones, kidney and bladder ailments, gastritis, and typhoid fever. Teas from the white eucalyptus leaves are used as a disinfectant for bathing open wounds, as a remedy for flu or colds, and when lightly brewed, a pleasant beverage. The little pods from which the blossoms grow at the tips of the branches are boiled with sugar to make an effective cough syrup, and in the past a strong tea made from new growth on the ends of the branches was widely taken by diabetics. For use as beverage, six or eight leaves are placed in a quart of water and boiled for three minutes, then steeped. When taken as a flu or cold remedy, the tea is made stronger and produces sweating.

Varying species of eucalyptus are also known as blue gum, sugar gum, gray gum, red gum, and white gum. It is important to note that the leaves, pods, and branch tips of only the white variety can be used in making medicinal teas.

All the other species of eucalyptus, or gum tree are commonly lumped together by the people of the Southwest and referred to as red. Tea cooked from the leaves of the red varieties will produce dizziness and severe headaches.

The only means that I know of distinguishing white eucalyptus from its many close cousins is by the whiteness of its trunk and limbs and by the fact that its leaves always appear green. Trunks and limbs of the red varieties, where the bark has shed, usually have a faint, greenish hue; and the foliage, when viewed from a distance, has a slight reddish color. But these factors can vary sufficiently so that it is still possible to mistake an extremely light colored red eucalyptus for a white one.

The flowers of white and red eucalypti are identical, and both varieties of trees grow both pink and white blossoms. The leaves of all eucalypti are dissimilar and cannot be used as a specific guide to identification. They may be short and oblong or long and pointed, but they all have the characteristic of growing unevenly on each side of the center veins. Because the trees shed their bark instead of their leaves, the leaves become tough and leathery. Though considerably fewer in number than other varieties, white eucalypti are always present in each grove and lane of gum trees. To the uninitiated, positive identification of the highly medicinal white eucalyptus is sometimes difficult. If a person is in doubt, he can determine with a single pot of tea which tree is the useful one for medicinal purposes.

Lumber sawed from eucalyptus is both hard and decorative. The trees make good windbreaks and shade. When used as firewood, eucalyptus should be sawed and split while it is green and can be worked easily. (*See photographs no. 74, 75.*)

Guatamote

Closely resembling willows in appearance, *guatamote* (pronounced goo-ah-tah-mo'-tay) grows along stream beds where water occasionally flows on the surface. Its stalks grow tall and straight and, in the past, have been used for arrows in spite of their pithy centers.

Mexican boys arm the tips of long *guatamote* shafts with cactus spines, and they harpoon or, rather, shoot, small edible fish in the clear streams of the Sierra Madre. They grasp the shaft in one hand as if it were an arrow, but in place of being hurled by a bowstring, it is instead propelled sharply forward by a flipping motion from a finger of the other hand. The cactus spines on the end of the shaft are frequently baited to attract small fish in the clear mountain streams.

In Mexico, where matches are taxed by the federal government and are not always readily available, Indians still use dry *guatamote* shafts for starting fires. The fire stick, pointed on one end and inserted into a shallow hole in another piece of dry wood, is speedily rotated back and forth between the palms until friction ignites the ample pith in the center of the stick. Once a spark is formed on the pointed end of the stick it is carefully breathed on and nurtured into a healthy glow which eventually bursts into flame.

As a diuretic, a cluster of leaves and branch tips of *guatamote* is pressed inside clothing in direct contact with the lower stomach where it remains until it brings relief. A tea is brewed from the leaves and small branches which quickly heals galled and infected feet; and fresh, young leaves are placed, and worn, inside socks and shoes to dispel foot odor.

Guatamote grows abundantly in southern California, parts of Arizona, and all of northern Mexico.

Crushed leaves mixed with olive oil are used to reduce swelling in glands. (*See photographs no. 76, 77.*)

Elderberry, or *Saúco*

Elderberry, or *saúco* (pronounced sah-oo′-co), trees are almost as much a part of the living Southwest as the *cholla*, mesquite, or greasewood. The trees grow along the banks of dry arroyos which carry water but briefly once a year and sometimes not that often. They intermittently dot the foothills of the desert, growing on sheltered slopes and in depressions where they are protected from the strongest winds.

Their pithy branches are weak, and they break easily, especially in springtime while the limbs are heavy with leaves and blossoms. The wood is soft and useless as a fuel, and the foliage is not sought after as a feed by livestock or game animals.

While elderberry is commonly thought of as a shrub, in the southwestern United States and in Mexico it grows indisputably as a tree. Its fruit forms in clusters and is easily picked or, rather, stripped by the handful from the branches. Sweet jams, jellies, and wines are rarely made anymore by the people who live in close proximity to elderberry, or *saúco,* but the medicinal properties contained in the leaves and flowers of the tree are still well known and widely used.

Teas of varying strengths cooked from the flowers are taken by expectant mothers for morning sickness and given to tiny babies for colic. In Mexico, two glasses of *flor de saúco* tea are still a standard dosage for breaking the fever of children suffering from measles. It is a soothing antacid tonic which, as a matter of custom, is sloshed liberally into burning stomachs the "morning after" by the participating members of gala southwestern *fiestas.* Hot tea is taken as a reliable cold and flu medicine, and *flor de saúco* is also boiled in milk to make a medicated cough syrup.

In Mexico an extract made from crushed elderberry leaves soaked in alcohol is taken in the proportion of ten drops to a glass of water for halting diarrhea, and tea cooked from its leaves is used for the treatment of dropsy. *(See photograph no. 65.)*

Gobernadora

Gobernadora (pronounced go-ber-nah-do'-rah), greasewood, and creosote bush are names frequently used for the same hardy shrub which grows unchecked for thousands of miles across the face of the great southwestern desert. Tea cooked from its leaves and branch tips is taken for coughs, colds, and many other afflictions including body odor. Poultices of the leaves are applied locally to relieve arthritic pain.

Steam baths from a weak solution of *gobernadora* tea produce much sweating and are taken as a treatment for flu, arthritic and rheumatic conditions, and as a general aid to personal cleanliness and good health. *Gobernadora* steam baths redden the skin and cause it to tingle vigorously. If continued they become increasingly painful, and discretion must be used in the number of baths given an ailing person. Native people of the Southwest recommend that the average individual should not have more than two *gobernadora* steam baths each year.

Its wood is easily ignited and makes a hot cooking fire which goes out quickly when it is no longer needed. *Gobernadora* shrubs have caused more flat tires for desert travelers than have all the sharp rocks, cacti, and other rough country hazards combined. The seasoned desert driver goes a long way around to avoid running a tire through the base of a *gobernadora* bush. (*See photographs no. 78, 79.*)

Higuera Cimarrona, the Wild Fig Tree of Baja

The *higuera cimarrona* (pronounced e-gay′-rah se-mar-ro′-nah) of Baja California is a strange-appearing tree with an odd inclination to grow only in impossible places. As its Spanish name implies, it grows alone and produces its fruit without companionship.

The literal translation of its name is wild fig tree, and excepting the miniature figs this eccentric tree produces, there is no similarity between it and the regular fig trees universally recognized for their succulent fruit and their historic prominence.

Higuera cimarronas, are usually found clinging precariously to the sides of windswept rocks high in the mountains of Baja, where their grotesque trunks and roots appear to be plastered across the faces of perpendicular ledges. A nonconforming *higuera cimarrona* will sometimes sprout on the ordinary slope of a hill instead of a sheer rock wall;

but, on these rare occasions, the tree still emerges from a crevice of dry rock. In the subsequent process of its retarded growth, the base of the tree spreads outward as though it were a thick, moveable substance, slowly enveloping as many rocks as it can grasp in its weird tentacles.

From central Baja California to Cape San Lucas these strange white trees, locally called *higuera cimarrona, zalate,* or *higuera silvestre,* grow at widely spaced intervals and always in the most difficult terrain they can find.

A tea boiled from the tree's large rounded leaves is taken by people in enormous quantities as an antidote for rattlesnake bites. Five-gallon cans of the same tea are brewed and poured down the throats of mules and cattle also suffering from the powerful venom of Baja rattlers.

Florentino Romero Jerardo, rancher, miner, and colorful character of Baja California for more than eighty years, was bitten by rattlesnakes on four different occasions. The first *picada,* as Spanish-speaking people call snakebite, was suffered by Florentino near La Paz. The second and third *picadas* occurred in the vicinity of San Bruno, a small village twenty miles south of Santa Rosalia, and the fourth happened near Bahía de los Angeles. In each instance Florentino's sole treatment after the snakebites were bled and sucked was the drinking of copious amounts of *higuera cimarrona* tea.

The fruit of the tree is exceptionally sweet when ripe, and the figs, though very small, grow abundantly.

The large *higuera cimarrona* standing near one of the original stone buildings of the Misión Santa Rosalia de Mulge complex is a splendid specimen and one of the first to be seen by southbound travelers following the Baja road. Residents of Mulge are proud of their ancient "snakebite" tree, which is close to the center of town, and they enjoy pointing it out to strangers and telling them of its virtues. Old-timers there say that if the tree has increased at all in size in the last seventy years, its growth is not noticeable. One Mulege rancher, now in his mid-seventies, has a small *higuera cimarrona* near his ranch house. He is especially fond of the tree and has watched it closely most of his life. The base has widened and now covers a slightly larger area on the rocky ridge where it stands, but the old rancher claims that the tree is still no taller than it was when he first saw it as a young man.

Farther down the peninsula, *higuera cimarronas* are encountered more frequently, and many of them, draped surrealistically over the sides of rocks and steep banks, can be seen from the road in the San Bartolo area south of La Paz.

Seasoned *higuera cimarrona* wood is prized by the natives of Baja California for making huge, durable bowls, called *bateas,* and kitchen cutting boards, which, like the living trees, seem to last unchanged forever. *(See photographs no. 80, 81, 82.)*

Dipúa

The *dipúa* (dee-poo′-ah) tree, which grows in Baja California, is much sought after by livestock when more natural feeds are short or non-existent. *Dipúa* closely resembles the well-known desert *palo verde,* but its branch tips are much finer. Bundles of *dipúa* limbs are customarily carried by packers and stockmen of Baja California. They are hacked from the trees with *machetes* and tightly bound with a thong. One bundle of limbs will feed an ordinary Baja California mule for at least twenty-four hours. Livestock, mules and burros especially, seem to relish the flavor of *dipúa,* and they can work hard on the strength derived from eating the limbs. *(See photograph no. 83.)*

Palo Fierro, or Desert Ironwood

The *palo fierro* (pronounced pah′-lo fe-er′-ro), or desert ironwood, tree provides the hottest-burning wood in the

82. Hundred-year-old *batea* made from *higuera cimarrona*

83. Bundle of *dipúa* carried by stockmen for feeding stock

84. Boys gathering wood for the Mexican village of San Ignacio

85. Iron-like *palo fierro* wood harvested with sledge hammers, not axes

86. Inéz Romero shaping a *palo fierro* black for harpooning

87. Pack-saddle blades made of lightweight cottonwood

88. Cottonwood cavalry barracks still standing at old Fort Cady

89. Leafy cottonwoods providing shade for summer siestas

desert. The dry, rust-colored snags remaining from the live trees of another era are so hard that it is almost impossible to cut them with any kind of a hand tool. Sledge hammers instead of axes are used for gathering the unbelievably heavy *palo fierro* wood, and the weathered pieces are pounded upon until they are loosened and can be pulled from the earth. The main roots never seem to decompose, and there is frequently more wood to be harvested below the surface than that which appears above ground.

In the past, *palo fierro* wood has been used extensively throughout the southwestern deserts for firing the boilers used in connection with mining operations, both large and small, and for making charcoal to fuel the blacksmiths' forges. Its properties of burning very hot and long make it ideal for these purposes. Today the wood is still very much in demand, and commercial truckers gather and sell it daily in many of Mexico's major cities. Approach any village in Baja California early in the morning, and you will encounter strings of pack burros being driven into the desert hills, usually by small boys, to be loaded with wood. These young wood gatherers collect principally a variety of dry cactus skeletons and mesquite, but whenever possible they concentrate on *palo fierro* because of its higher market value in the pueblo.

The mature *palo fierro* trees, which frequently grow to twenty feet, spread out nicely, providing good shade. The branches have intermittent, hook-shaped thorns. The young *palo fierros*, which resemble shrubs more than trees, are called *uña de gato* (catclaw) by local people.

When an extremely durable piece of wood is needed in Baja California for a mechanical bearing or the shaft of a turtle harpoon, a piece of green, unchecked *palo fierro* is cut and carefully whittled into whatever shape is required to do the specific job. For a turtle harpoon, a head of steel can be fitted over a tapered *palo fierro* block. Rubber washers are then placed near the barbed head to prevent its penetrating too deeply and injuring the turtle, as the green sea turtle found in the Gulf of California must be kept alive until marketed, often two to three weeks. (*See photographs no. 85, 86.*)

Alamo, or Cottonwood

The person who has never stretched out on the ground and pillowed his head against the trunk of an *alamo* (pronounced ah-lah-mo), or cottonwood tree, for a summer siesta has missed one of life's most gratifying experiences.

The cottonwoods of southern California's desert lands provide the nicest settings of all for pleasant siestas. For thirty years I've been dozing delightfully under cottonwoods from Wyoming to Mexico, and few people are better qualified to comment on this supine pastime than I. To appreciate fully the fine art of taking summer siestas, strangers and newcomers to afternoon napping out-of-doors might profit by pondering a few authoritative words on the subject.

First, they should realize that California's cottonwoods are especially friendly and that they grow in the most comfortable places. They require more water than native trees like pines and oaks, so therefore are usually found near springs or growing along the banks of clean stream beds which run water, at least periodically, each year.

To go about a summer siesta properly, you must lie quietly on your back so you can look up through the leaves at the sky. Next, you begin to concentrate on the message that one branch of rustling leaves is sending to a reciprocating cluster of quivering foliage on the far side of the tree. Cottonwood leaves are always chattering, even though the air is still, and their gossip goes on and on. They whisper in soothing tones about abstract matters like the care and feeding of the newly hatched birds on a certain limb or the silently changing cloud formations high above.

Select a position under the tree where you will have constant shade during the timeless period of your siesta and one where your presence does not detract measurably from the ants' urgent business matters. Ants always have a main thoroughfare over which they travel across the ground and up and down the trunks of cottonwoods. If a person takes the time to avoid this mainstream of activity, especially if he's a beginner, he can usually sleep longer and more soundly.

For the ultimate effect, prepare for your siesta *only* after you have completed lunch, and then carefully proceed with deliberate, unhurried movements. If a slight breeze is blowing, it becomes harder to follow the conversation flying back and forth between the fluttering, muttering leaves, but sleep generally comes more quickly this way. At that time of year when the cottonwoods are dispelling their snowy airborne seeds, the seasoned siesta specialist, to prevent the cotton from tickling his nose, breathes from the crook of an arm thrown limply across his face.

With just the least amount of effort, anyone can soon determine that a cottonwood siesta is equally as relaxing as a sauna, and it's so much less trouble.

Besides inducing a joyful sleep at siesta time, cottonwoods also have definite medicinal values. Indian and Mexican people of the Southwest boil a strong disinfectant tea from its leaves and branch tips, and they apply it to ulcerated lesions and infected wounds.

The wood itself had other uses besides that of fuel. Because of its extremely light weight when dry, it was once used extensively for ox yokes (not to be confused with the hardwood *bows* which encircled the oxen's necks and held the yokes in place). Among saddlemakers, it is still a popular wood for making blades for the trees of pack and riding saddles.

The U.S. cavalry barracks of old Fort Cady on the Mojave River, east of Yermo, California, are made of cottonwood logs. The cavalry moved out of Fort Cady in 1882 after the government decided that wagon-train immigrants were no longer in danger of Indian attacks, but the buildings there are intact and usable. One end of the barracks, after having been occupied for as long as anyone can recall, is still being lived in at this writing by a bachelor who works on the Cady Ranch.

The frequent occurrence of the Spanish words *Alamo, Los Alamos, Alameda, Alamogordo,* and *Alamitos* on maps of the Southwest is by no means coincidental. Cottonwoods thrive naturally in nearly all our states; but where they find water in the warmer climes of the Southwest, they grow to giants of tremendous size and beauty, live for at least a century, and well deserve the honor of having so many canyons, mountains, springs, ranches, and villages named for them.

Cottonwoods were regularly planted by early Californians to provide shade around ranch dwellings or to make shade for livestock near windmills or barnyards. The old-timers, as a matter of convenience, used fence posts made of cottonwood, and in bottom lands where the earth was moist, an unbelievable percentage of them would sprout, take root, and grow into huge leafy shelters along the fence lines where many still remain.

The chief enemy of southern California's cottonwoods is mistletoe. Whenever the parasitic plant is allowed to grow unchecked through the branches of these stately trees, it quickly sucks the life out of them. Then their grotesque skeletons soon topple, and the huge landmarks turn full cycle by settling back into the earth. *(See photographs no. 87, 88, 89.)*